Bridging Circuits and Fields
Foundational Questions in Power Theory

Alexander I. Petroianu
University of Cape Town, Cape Town,
Republic of South Africa
University of Calgary, Calgary, Alberta, Canada

CRC Press
Taylor & Francis Group
Boca Raton London New York

CRC Press is an imprint of the
Taylor & Francis Group, an **informa** business

A SCIENCE PUBLISHERS BOOK

First edition published 2022
by CRC Press
6000 Broken Sound Parkway NW, Suite 300, Boca Raton, FL 33487-2742

and by CRC Press
2 Park Square, Milton Park, Abingdon, Oxon, OX14 4RN

ISBN: 978-1-138-71043-6 (hbk)
ISBN: 978-0-367-71130-6 (pbk)
ISBN: 978-1-315-20067-5 (ebk)

DOI: 10.1201/b22123

Typeset in Palatino Roman
by Innovative Processors

Acknowledgements

During this long writing period (almost four years), I have been a 'present absence' for my family; I appreciate their love and support (although some, including me, thought that I would never finish it).

This is also a moment to express my gratitude to the University of Cape Town where I have served since 1989; this is one of the oldest universities in the southern hemisphere. I was able to share some of the ideas on this manuscript with my colleagues and students and to use rare documents in their library related to the beginning of electricity.

I express my gratitude towards the University of Calgary, where I served as an adjunct professor for many years. The university library is one of the most advanced in the Northern Hemisphere. The online access to digitized sources (especially in these difficult times when libraries are closed) enabled me to finish the monograph.

Living in Canada has given me the peace of mind necessary to continue a life with purpose.

A special 'thank you' to Luisa, my wife, love, and friend, and one of the most severe copy-editors I have ever met – she reminded me permanently of Boileau's dictum:

> *Ce que l'on conçoit bien s'énonce clairement,*
> *Et les mots pour le dire arrivent aisément.*[1]

With her sharp criticism and helpful suggestions, she turned my idiosyncratic and terse engineering text into a more readable manuscript. More than that, without her, I would never have finished the monograph! Thanks also to Sylvia Izzo Hunter, who helped to revise the bibliography.

As a last thought: I have written this monograph also in loving memory of my parents (Iosif and Evelyne) and of my first wife, Susanne Petroianu *née* Bock.

---

[1] What is clearly thought out is clearly expressed.

Preface

Motto: To those who came before us and those who will come after us searching for the elusive meaning of electrical power.

Dedicated to Charles (Karl) Proteus Steinmetz – with admiration and respectful dissent.

If it is true that professors never retire, at the most they lose their faculties, then I missed the second option and, many, many years ago started to investigate (paraphrasing Riemann) the hypotheses underlying the concept of power in electrical circuits.

Starting from the mundane power plug, the search for the meaning of power followed a long, tortuous, but fascinating journey, which started from the beginning of electricity and electromagnetism and continued up to today's quantum mechanics and relativity.

No discipline can be understood without knowing its past. The monograph covers many of the historical landmark writings on electrical and electromagnetic sciences, uncovers the filiation of ideas and pays due recognition to the founding fathers' heritage.

At macro level, power engineering deals with voltage, current and power; at atomic or mesoscopic level, power engineering deals with electrons, positrons, photons and Planck's quanta; at this level physics meets metaphysics or philosophy of science.

The monograph reveals not only the many mathematical avatars of the concept of power, but also the limitations that the current state of physics imposes on us and on our understanding of the processes related to power transmission phenomena at the mesoscopic level.

This monograph is written by an engineer for his colleagues – power engineers. The mathematical formulae and expressions are kept to a minimum. However, this does not make the monograph easy reading. This is not a book that thinks for you; this is a book that should make you think about a very serious issue – power engineering as a discipline is still lacking a theory of power!

As power engineers, we are like the proverbial shoemaker whose children have no shoes. We dedicate our professional life to a scientific discipline – power engineering – without having a clear idea of what we are dealing with.

The monograph addresses this question.

Chapter 1 presents my motivation for writing the monograph, and why this subject matter is important to a large audience.

Chapter 2 gives a short historical overview of the genesis of power theory. I was lucky to stumble over a mathematical failure of Steinmetz – one of the founding fathers of electrical engineering – and, as a result, I, the author, was confronted with a paradox – not Steinmetz! How could a mathematically incorrect equation used by Steinmetz in his symbolic method give correct results? And how could it be that, for the same concept – power – we have two different mathematical descriptions (from Steinmetz and Janet)? This chapter also reveals the limitations of the existing paradigm. Power theory is only a mathematical theory: it says nothing about the physical structure of electric power and does not reveal the mechanism of power transmission.

Chapter 3 takes a position on the Czarnecki-Emanuel debate about the relevance of the Poynting Theorem for power theory and shows that this debate has historical roots in the older debate between Abraham and Minkowski about electromagnetic momentum.

Chapter 4 presents a new power paradigm that conforms to the latest developments in physics and is expressed in the language of geometric algebra. The chapter proposes both a new mathematical expression for electromagnetic power and a new physical interpretation of power.

Chapter 5 gives an overview of the many mathematical 'guises and disguises' by which the power concept has been presented in the literature.

Chapter 6 concludes that power theory is a mathematical theory only and remains unfinished from the physical point of view. I conclude that power theory is still a growing discipline. As Heaviside said, "There is no finality in a growing science."

Alexander I. Petroianu

Contents

Introduction

1. The Subject Matter: Why Does it Matter?

The subject of the monograph is *power* in electrical circuits and transmission networks. The author examines the many mathematical representations of the electrical power concept.

Electrical power is the subject of power theory, which is an important discipline in electrical engineering. No discipline can be successfully investigated without knowing its history; for this reason, the monograph reflects the development of the power concept in its historical context. The development of power theory is tightly connected with the history of science and developments of two important schools of thought: Continental Electrodynamics and Faraday-Maxwell Electromagnetism.

Power phenomena occur over a broad spatial and temporal range. At their smallest scale, they approach Planck's quanta, and at their largest scale, their velocity approaches the speed of light. At these limits, physics meets metaphysics. The conceptual problems of power theory overlap the fundamental problems in the philosophy of science.

1.1 Author's Motivation

As a student, the author did not understand *why*, in direct current (DC) circuits, electrical magnitudes are represented as real numbers, whereas in alternating current (AC) circuits electrical magnitudes are represented as vectors, phasors, and complex numbers.

As an engineer involved in power system operation and control, the author focused on large-scale computer applications such as power flow and state estimation; this did not leave too much time for thinking about the meaning of the terms active or reactive power. However, he had a nagging feeling that double-frequency power components oscillating between source and load contradicted the principle of least action.

As a teacher, the author had difficulty explaining to students how a "crawling" electron (with speed through the conductor of approximately 10^{-4} m/s) enables an almost instantaneous transfer of energy, over long distances, from generator to load (the velocity of energy transfer is near to the velocity of light in vacuum, approximately 3×10^5 m/s).

In the context of electromagnetic theory, students are taught, on the basis of Poynting's theorem, that energy flows from outside the wires into the wires. But when studying circuit theory, they are taught, on the basis of Kirchhoff's law, that currents of electrons carrying power flow through the wires. Teaching the same concepts (e.g., power flow, energy flow) in completely different ways is inconsistent and confusing to the students.

Teaching something that the teacher himself does not understand reminds us of Faust (Goethe, 1790), who frustrated with his inability to understand the world, turns to magic:

> Dass ich nicht mehr mit saurem Schweiss
> Zu sagen brauche, was ich nicht weiss.
> So that no more with bitter sweat
> I need to talk of what I don't know yet

(translation by George Madison Priest) http://userhome.brooklyn.cuny.edu/anthro/jbeatty/COURSES/German/german1.2/faust.html

As academic fields, circuit theory and electromagnetism are closely related; however, they are separated by a huge gap with regard to their mathematical formalism and physical interpretation of the same fundamental concepts: power and energy. In other words, each of these disciplines is taught in "splendid isolation."

At the end of a long career in power engineering, the author realized that he is not alone in being far from understanding the question: *what is electrical power?* Even an eminent physicist like Feynman confessed that he did not understand what energy is.

Satisfying the author's curiosity and his frustration with old vexing problems perhaps justifies writing this monograph. However, this is not an argument for you, the reader, to read it. Here are the arguments showing that the question *"what is power?"* should bother you as well.

1.2 Reader's Motivation

For more than a century, the power industry has successfully generated, transmitted, and distributed huge quantities of power over long distances. The question that could be asked is, if it works and nothing is broken, what should we worry about, and why? The answer is that we have two types of power engineering knowledge: procedural and conceptual. This monograph deals only with conceptual knowledge. Its scope is not to change or improve how we calculate load flow or state estimation, but rather to understand what we are calculating: what physical and mathematical entities we are dealing with. The goal is to gain insight rather than simply to follow Leibniz's exhortation "calculemus" (let us calculate).

1.3 What is Electrical Power?

This is a simple, obvious question for everybody using power. Sometimes, the most obvious and simple questions are the most difficult ones.

It took the author 14 years to ponder this question and more than four years to write down an answer to the following questions: What is electrical power? How much do we understand what we know, where the limits of our knowledge are, how to represent the concept of power mathematically, and how to interpret it physically?

Contrary to opinions such as: "The concepts and definitions of electric power for sinusoidal AC systems are well established and accepted worldwide" (Akagi et al., 2007:19), there are still numerous open questions related to power theory.

Emanuel (2010:xiii) complains about the "inadequate power definitions" and states that "a lively debate over the apparent power definition and its resolution started a century ago and has not yet reached a conclusion."

Czarnecki (1997:360) states that "interpretation of power phenomena of circuits with nonsinusoidal voltages and currents remained unclear and controversial."

Küpfmüller et al., in their book Theoretische Elektrotechnik: Eine Einführung (2013:44), state that there is no consistent mathematical representation of the AC calculation from Steinmetz and Kennelly.

The mathematical analysis of Steinmetz's symbolic method (Someda, 2006) reveals that his definition of active power as an inner product of two vectors is different from the definition of power as the product of two complex numbers. The difficulty of matching the two expressions resides in the fact that vector space and complex space are not isomorphic spaces.

Atabekov (1965) considers complex power to be a phasor quantity, which can be represented graphically in a complex space, whereas Faria (2008:260) emphasizes that "complex power is not a phasor."

The author tends to agree with Kennelly (1910:1265), one of the founding fathers of electrical engineering, who stated more than one hundred years ago that "the algebra and geometry of vector alternating current technology...are, at present, in a state of great and unnecessary confusion....[which] has existed for more than twenty years and is not confined to any one country or language."

These examples (and there are many more) show that the mathematical and physical foundations of the power concept are in need of revision.

In an anecdote associated with Faraday and electricity, it is said that when Gladstone, the British prime minister, asked Faraday what use his discovery of electromagnetism might have, Faraday replied: "Someday you will be able to tax it." (Cohen, 1987: 177-182). To the reader's putative question about the worth of reading this monograph, the author's answer is that perhaps in reading it, you might understand what are you taxed for.

The question "What is electrical power?" is a matter of interest for students, practitioners, and specialists in electrical and power engineering, electrodynamics, electromagnetism, signal theory, physics, and applied mathematics.

2. Foundational Issues Related to the Concept of Electrical Power

2.1 Ontological Point of View

The author examines the physical meaning of entities such as voltage, current, and active, reactive, and apparent power, as well as how these entities are reflected in classical electromagnetic theory, or more specifically, how these entities are related to force, energy, and momentum.

2.2 Epistemological Point of View

The author examines the different mathematical formalisms (real algebra, complex algebra, trigonometric algebra, vector and tensor calculus, geometric algebra, etc.) used to represent physical entities in circuit theory and in electromagnetism.

The concept of electrical power is not antecedent-free; it is related to the concept of force. The monograph shows that the Blakesley-Ferraris expression for electrical power derives from the Amperian concept of force. The author stresses the important contribution of the Continental Electrodynamics school (Coulomb, Ampère, Poisson, Gauss, Riemann, Grassmann, Neumann C. and Neumann F., Weber and Poincaré) to the development of the expression for power: $p = vi$.

In Germany, one of the strongest supporters of the Faraday-Maxwell electrodynamic theory was August Föppl, whose book was cited by Steinmetz (as well as by Einstein). Föppl influenced Steinmetz's approach to power theory in two ways: 1) by introducing vectors as geometric representations of electrical magnitudes and 2) by his skeptical attitude toward the physical reality of voltage and current.

In recent years, there was a heated debate between Czarnecki and Emanuel related to the relationship between the Poynting vector and the expression of electrical power. For this reason, the author investigated the large body of literature related to electromagnetism (Faraday, Maxwell, Heaviside, Hertz, Lorentz, Lorenz) and electrodynamics (Ampère, Gauss, Weber, Neumann, Poincaré).

As a result of developments in physics (such as the Aharonov-Bohm effect, gauge theory, and the Yang-Mills experiment), electromagnetic theory is still being re-evaluated, and alternative theories have been proposed by writers such as Liènard-Wiechert, Wheeler-Feynman, Born-Infeld, and Hartenstein. However, the schism between quantum theory (Dirac, Heisenberg, and Schrödinger) and relativity (Einstein) remains; this is the reason why electromagnetic theory remains an unfinished theory as well. Because of the existing "crisis" in physics, many fundamental concepts (including the concepts of energy and power) are still under debate.

3. Contributions of this Monograph to Power Theory

3.1 Reappraisal and Reformulation of Steinmetz's Symbolic Method

The author demonstrates that Steinmetz was "wrongly right": his expressions for active and reactive power are correct, although his mathematical "proof" is false. Against his conviction, Steinmetz did not use complex algebra or vector calculus. He rediscovered Grassmann-Clifford Geometric Algebra. Unfortunately, he fooled not only himself, but many, many generations of electrical engineers who used geometric algebra thinking that they were using complex algebra and vector calculus.

3.2 Reappraisal of Janet's Heuristic Expression $S = \dot{V} I^{*}$

The author through geometrical reasoning demonstrates that Janet's rule ensures the invariance of two geometric areas, one representing active power and the other, reactive power. The author demonstrates that Janet was also right, only he did not know *why!*

3.3 Demonstration of the Mathematical Isomorphism between Steinmetz's Power Expression and Poynting's Expression for Energy Flow

The author demonstrates the mathematical isomorphism between Steinmetz's power expression and Poynting's expression for energy flow. In a similar way, it is shown (as many other authors have shown before) that Janet's expression is mathematically isomorphic with Poynting's expression for energy flow.

What differentiates this monograph from previous contributions (as reflected in the Czarnecki-Emanuel debate) is the fact that it contradicts the ontological interpretation of all the other authors. Behind a formal mathematical isomorphism (the epistemological aspect) hides a fundamental ontological contradiction. The author rejects, on the basis of physical interpretation, the relevance of the Poynting vector for power transfer in electrical circuits. A mathematical proof does not automatically represent a physical truth. The author rejects the idea that power comes from the exterior electromagnetic field and compensates for the Joule losses. The author rejects the idea that electrical networks are merely passive antennas. The interaction between the electromagnetic field and the electrical charges of the conductor are more complex than Poynting's interpretation. Power transfer between generator (sources) and the load (sinks) involve both electromagnetic waves and electrically charged particles of the conductor. This monograph offers a hypothesis of helicoidal-like power transfer (a combination of linear and rotational momentum).

3.4 Reactive Power is as much Power as Active Power

The author considers that there is only *one* power: *active* power is a tautology, and *reactive* power is a contradiction in terms (an oxymoron, or at least a misnomer). Reactive power is as much power as active power. One could be considered as derived from potential energy, and the other, from kinetic energy. *Reactive* power is in fact a momentum and a condition *sine qua non* (something absolutely indispensable or essential) for the transfer of potential energy (or power). The "struggle" for reactive power compensation is an ill-posed problem. We need momentum (linear in DC circuits and linear plus rotational in AC circuits) in order to move/transport/transfer energy from generators to the consumers. Momentum represents the mechanism for power transfer in both DC and AC circuits; for this reason, is incorrect to link the concept of *reactive* power with only the AC circuits.

3.5 Apparent Power does have Physical Meaning

Authors in the majority of power engineering publications consider that apparent power has no physical meaning: the KVA (kilovolt ampere) is just a "design" magnitude useful to calculate the dimensions of electrical machines, transformers, and apparatuses.

The author contradicts this opinion. Apparent power is equivalent to Lorentz's force in classical electromagnetic theory. The author gives a new interpretation of the power expression:

$$S = VI = V \cdot I + V \wedge I$$

This expression represents (in the mathematical formalism of geometric algebra) the equality between an external force and the sum of electromagnetic energy and momentum (in a closed power system).

3.6 Criticism of the Interpretation of Double-frequency Terms

The monograph criticizes the interpretation of the double-frequency terms that appear as result of multiplying voltage by current in their trigonometric representations; the majority of textbooks interpret these terms as *oscillations* of power between generators and loads, or as power (active and reactive) surging back and forth, with no net energy transfer. The author challenges this interpretation, which contradicts the principle of least action and resembles a kind of (energetic) perpetuum mobile within the power network.

3.7 Validity of the Instantaneous Power Concept

The author questions the validity of the *instantaneous power* concept. The idea of instantaneous power obtained as a measurement of instantaneous voltage and instantaneous current contradicts Heisenberg's uncertainty principle in quantum mechanics.

3.8 Physical Interpretation of Voltage and Current as Inseparable Entities

The author questions the physical interpretation of voltage and current as separate or separable entities. The author interprets voltage and current as two mathematical representations or faces of the same physical entity: *electromagnetic power*.

3.9 Issues Related to Load Flow and State Estimation

The author raises some issues related to load flow and state estimation. These analytical tools were developed within the framework of "classical" power theory. Classical power theory, likewise classical electromagnetism, considers the *scalar electric potential* but does not consider the existence of the *vector electromagnetic potential*. It also ignores the delayed and action-at-a-distance (aaad) electromagnetic effects. The development of quantum mechanics and mathematics proves the need to re-assess the existing analytical tools currently in use for power system analysis, operation and control.

4. Research Methodology

Power theory, as a sub-discipline of electrical engineering, is at the crossroads of Physics and Mathematics; it is also tightly interconnected and influenced by other disciplines, such as circuit theory, electrodynamics and electromagnetism, signal theory, electrical machines and power electronics, quantum theory, relativity, algebra, and geometry.

For this reason, the author adopted a multidisciplinary approach in his investigation of the foundational concepts related to electrical power.

This approach required a broad knowledge of both the literature on many topics and the development of these topics over time. As a result, the author sometimes trespasses into the domain of the history of electrodynamics and electromagnetism. The monograph also deals with epistemological and ontological topics pertaining to the philosophy of science that, although beyond the scope of a strictly technical investigation, are highly relevant for a proper understanding of the power concept.

5. Literature and References

The author examined a large body of primary and secondary literature pertaining to the evolution of the power concept. Special attention has been given to the heritage left by the founding fathers of electricity, electrodynamics, and electromagnetism. The author tried to determine the filiation of ideas and the interconnectedness of electrical engineering, electromagnetism, physics, and mathematics with regard to concepts of power, force, energy and momentum.

The author also consulted the following databases: EBSCO Academic Search, Archives-Ouvertes, ArXiv.org e-Print Archive, Compendex, DART-Europe E-Theses, Eidgenössische Technische Hochschule (ETH) Library (Zürich), Elibrary.RU (Russia), École Polytechnique de l'Université de Lausanne (EPFL) Theses, Gallica Digital Library, Google Scholar, Hathi Trust Digital Library, Göttinger Digitalisierungszentrum (GDZ), IEEE Xplore® Digital Library, Inspec, JSTOR, Knovel, MathSciNet, Project Gutenberg, ProQuest Dissertations & Theses, ScienceDirect, Scopus (Elsevier), SPIE Digital Library, Springer Link, Technische Informationsbibliothek (TIB) – German National Library of Science and Technology.

The research for this monograph required access to books, archives, and online resources. The author has been fortunate to belong to the University of Cape Town, one of the oldest in the Southern Hemisphere (unfortunately destroyed in a devastating fire). Its archives contain a treasure trove of telegraph and electricity journals from the early nineteenth century.

He also greatly benefited from his affiliation to the University of Calgary; its library is one of the most modern in North America with regard to online access to books, publications, and databases in engineering, physics, and mathematics.

6. Style

Since the monograph is focused on the conceptual aspects of the power theory, mathematical formalism is kept at a minimum. However, readers need a basic knowledge of algebra (real-, complex-, trigonometric algebra) and geometry (vector calculus and geometric algebra).

7. Structure

In Chapter 1, the author presents his motivation for writing the monograph, and why this subject matter is important to a large audience.

Chapter 2 gives a short historical overview of the genesis of power theory. The author was "lucky" to stumble over a mathematical failure of Steinmetz – one of the founding fathers of electrical engineering – and, as a result, he was confronted with a paradox. How could a mathematically incorrect operation (Steinmetz's symbolic method) give correct results? And how could it be that, for the same concept – power – we have two different mathematical descriptions (from Steinmetz and Janet)? This chapter also reveals the limitations of the existing paradigm. Power theory is only a mathematical theory: it says nothing about the physical structure of electric power and does not reveal the mechanism of power transmission.

In Chapter 3, the author takes a position on the Czarnecki-Emanuel debate about the relevance of the Poynting Theorem for power theory and shows that this debate has historical roots in the older debate between Abraham and Minkowski about electromagnetic momentum.

In Chapter 4, the author presents a new power theory paradigm that conforms to the latest developments in physics; this paradigm is formulated in the mathematical language of geometric algebra. The chapter proposes both a new mathematical expression for electromagnetic power and a new physical interpretation of power.

Chapter 5 gives an overview of the many mathematical "guises and disguises" by which the power concept has been presented in the literature.

Chapter 6 concludes that power theory is a mathematical theory only and remains unfinished from the physical point of view. The author concludes that power theory is still a growing discipline. As Heaviside said, "there is no finality in a growing science."

Power Theory in Electrical Circuits

1. Introduction

Defining the concept of power means opening a can of 'scientific worms'.
Griffiths (2013:357) defines power as the rate of *work*:

$$\text{Power} = \frac{dW}{dt} = \int_v (E \cdot J)dt$$

where work or energy $E \cdot J$ is deployed in per unit (p.u.) *time and volume* (i.e., energy has both time and space attributes).

However, Lehrman wrote a paper entitled "Energy is Not the Ability to Do Work" (Lehrman, 1973).

Valkenburg (1964) defines power as only a *time*-derivative of energy:

$$p = \frac{dw}{dt}$$

He factorizes the expression for the time derivative of *energy* into a product of voltage and current:

$$\frac{dw}{dt} = \frac{dw}{dq} \times \frac{dq}{dt} = v \cdot i$$

Einstein's definition of energy

$$E = mc^2$$

has long been accepted, but according to Okun (1989, 2006), it is incorrect.

Moreover, the concept of *energy* is related to the concept of *mass*, and *mass* itself is a 'messy' concept, as discussed by Psarros (1996) in his article, entitled, 'The Mess with the Mass Terms'.

No wonder that famous scientists like Feynman (1961) declare that physicists do not know what energy is.

The concept of energy, born in a period spanning the years 1800-1830 (Coopersmith, 2010), remains a subtle, elusive, unobservable, and evolving concept.

And what is true for the concept of energy is also true for the concept of electrical power. Power remains a subtle, elusive, unobservable, and evolving concept.

2. A Critical Assessment of the Existing Power Paradigm

The current paradigm expresses power in four mathematical forms: 1) power as a real-valued function in real algebra, 2) power as a complex-valued function in the algebra of complex numbers, 3) power as a trigonometric expression in trigonometric algebra, and 4) power as a vector-valued function in vector calculus.

In DC circuits, power is a real-valued function expressed as: $p = vi$.

In AC circuits, the current paradigm, based on Steinmetz's symbolic method, is an entanglement of three different mathematical formalisms:1) trigonometric algebra, 2) complex algebra, and 3) vector calculus.

In addition, Janet's expression for complex power: $\dot{S} = \dot{V}I^*$ is mistakenly considered to be equivalent to Steinmetz's formulation.

2.1 Steinmetz's Assumptions Underpinning His Symbolic Method: A Critical Review

In 1890 and 1891, Steinmetz introduced the representations of electrical magnitudes as vectors in polar coordinates; he mentions Blakesley, Föppl, and Kapp as predecessors in using vectorial representations (Steinmetz, 1890, 1891).

In his communication at the International Electrical Congress, Steinmetz proposed the representation of electrical magnitudes as complex variables (Steinmetz, 1893). Therefore, he introduced a double mathematical representation: 1) electrical magnitudes as *algebraic* complex numbers (in rectangular coordinates), and 2) electrical magnitudes as *geometric* (polar) vectors.

In his book, Steinmetz (1897) brings new terms, concepts, and ideas such as:

- The terms *reactive power* and *apparent power*
- The idea that the power factor could differ from unity, even if the voltage and current are in phase
- The idea that electrical power is a wave of *double frequency*

In a paper published in 1899, Steinmetz expresses power as $P = [\dot{E}\dot{I}]$, in which the square brackets symbol $[\ldots]$ "*denotes the transfer from the frequency $\dot{E}$ and $\dot{I}$ to the double frequency of P.*" His expression implies that the vector diagram for power should be separated from the vector diagram for voltage and current.

In the same paper, he gives a new interpretation of the symbol 'j' as follows:

Since $j^2 = -1$, that is, 180° rotation for E and I, for the double frequency vector P, $j^2 = +1$, or 360° rotation.

He further states that for the double frequency calculations,

$$j \times 1 = j$$
$$1 \times j = -j$$

(Steinmetz 1899:270).

He adds that "the introduction of the double frequency vector product $P = [\dot{E}\dot{I}]$ brings us outside of the limits of algebra...and the commutative principle of algebra: $a \times b = b \times a$ does not apply any more... [EI] unlike [IE]."

On the basis of the above conjectures, he derived the expression of power in rectangular coordinates:

$$P = [EI] = [EI]^1 + j[EI]^j = (e'\, i' + e''\, i'') + j(e''\, i' - e'\, i'')$$

(Steinmetz 1899:272).

Steinmetz's books published in German (Steinmetz, 1900a) and French (Steinmetz, 1912) repeated the main ideas of the 1899 paper and ensured the wide diffusion of the symbolic method in Europe.

The author of this monograph stumbled over Steinmetz's assertion that for double frequency (2ω), the square of the imaginary j is equal to plus one: $j^2 = +1$. It was obvious that Steinmetz had committed a mathematical blunder, and nothing is so exciting as to discover a failure in the thinking of an important scientist. Steinmetz was an accomplished mathematician, although with an unfinished mathematical dissertation at the University of Breslau. He obtained an additional engineering degree from ETH Zürich. It was surprising for somebody of this stature to state that $\sqrt{-1}^2 = +1$, which contradicts a basic tenet of complex algebra.

For this writer, this discovery represented a stumbling stone, the end of the symbolic method orthodoxy, and the beginning of inquiry into the validity of the symbolic method and the existing power theory paradigm. The path of his subsequent research, and of this monograph, was similar to opening a set of Russian dolls – a riddle, wrapped in a mystery, wrapped in an enigma, wrapped in a puzzle, wrapped in a conundrum, wrapped ...The riddle: if Steinmetz is wrong, why is his method heuristically right? The mystery: if Steinmetz is right, how could it be that Janet is also right? The puzzle: could it be that both are wrong, or both are right? The conundrum: could we have a plurality of mathematical expressions for the same physical entity?

Steinmetz's symbolic method has long been the subject of various debates and criticisms.

Kennelly (1893) supported the use of complex variables to represent electrical magnitudes, whereas Franklin (1903) considered that complex algebra is insufficient for representing the concept of power.

Kafka (1925) stated that a complex representation 'hides' the essence of the physical phenomena and considered that the electrical magnitudes should be represented as planar vectors.

Natalis (1924) supported a purely geometric representation, whereas Creedy (1910) proposed a representation of electrical magnitudes based on Möbius algebra.

Punga (1901, 1938) proposed to use, instead of the algebraic product, Grassmann's geometric product; in fact, he was the first scientist to promote Grassmann algebra in electrical engineering, before Bolinder (1959).

Perrine (1897) stated that complex numbers are not vectors. Besso (1900), a close collaborator of Einstein, criticized Steinmetz for not distinguishing between vector multiplication and complex multiplication. Emde (1901) criticized Steinmetz for confusing vectors with complex numbers.

Patterson (1911) stated that treating harmonic quantities as vector quantities gives wrong results in multiplication. Pomey (1918) stressed the difference between the commutative rule of multiplication in complex algebra and the non-commutative rule of vector multiplication in Gibbs' vector calculus.

Kennelly and Valander (1919) and Sah Pen-Tung (1936) also criticized Steinmetz for confusing complex numbers with vectors.

Oberdorfer (1929) considered that the use of planar vectors confines the analysis to a two-dimensional space.

However, even an influential textbook authored by M.I.T. professors as late as 1948 stated that there is "a common engineering use of the term *vector* interchangeably with complex quantity" (E.E. Staff, M.I.T., 1948: 266).

Despite its theoretical inconsistencies, Steinmetz's symbolic method entered the arsenal of analytical tools as an efficient use of complex numbers for power calculations in AC circuits. It is known under two expressions:

Steinmetz's expression for power:

$$P = [EI] = [EI]^1 + j[EI]^j = (e'\, i' + e''\, i'') + j(e''\, i' - e'\, i'')$$

and Janet's expression for complex power:

$$\dot{S} = \dot{V}I^*$$

The symbolic method's resilience is based on Steinmetz's huge scientific authority and the weight of authoritative support he received from his peers, such as A.E. Kennelly, P. Janet, C.-F. Guilbert, J.L. La Cour, and O.S. Bragstad. This resilience could be an interesting topic for the sociology or psychology of science.

The author considers that:

- The symbolic method is mathematically inconsistent because it relies on two irreconcilable structures: 1) complex algebra (a commutative algebra) and 2) vector calculus (which is not an algebra).
- The symbolic method considers that electrical magnitudes, from an algebraic point of view, are complex numbers. Complex algebra imposes a limit on the dimensionality of its entities; complex numbers exist only in a two-dimensional (2-D) space.
- The symbolic method considers that electrical magnitudes, from a geometric point of view, are vectors. Vector calculus is axiomatically endowed with two types of multiplication: the inner product and the outer product. The inner or dot product of two vectors results in a scalar. The outer or vector product of two (polar) vectors results in an axial vector. Vector calculus is not a closed set; each type of multiplication results in an entity different from the entities that were multiplied. Vector calculus is not an algebra.
- Vector calculus imposes a limit on the dimensionality of its entities. If voltage and current are considered planar vectors, the result of a vector or outer multiplication is a vector perpendicular to the plane in a third dimension (i.e. it contradicts the assumptions of the two-dimensional representation). In addition, vector calculus permits vector multiplication only in three-dimensional and seven-dimensional spaces (Eckmann, 1943, 2006; Gogberasvili, 2005; Silagadze, 2002). The vector product cannot be extended to spaces of other dimensions.
- The failure of Steinmetz's symbolic method is rooted in his adherence to Peacock's principle of permanence. This principle states that algebra should follow the rules of arithmetic, and thus, the multiplication obeys the rule of commutativity, i.e.,

$$ab = ba$$

With regard to the expression $\dot{S} = \dot{V}I^*$ proposed by Janet, there is no mathematical proof why the canonical and axiomatic definition of power $p = vi$ should be rewritten in complex as $\dot{V}I^*$.

The author agrees with the dictum of Steinmetz (1911:540): "In some fields of electrical engineering or of electrical science, we might almost say that we know less now than we knew, or rather believed we knew, a quarter of a century ago. There are things which had been investigated a quarter of a century ago and which were explained in a satisfactory manner to our limited knowledge in the early days, but this explanation does not seem satisfactory now, with our greater knowledge."

In this monograph, the author re-assesses and rigorizes Steinmetz's symbolic method and gives a mathematical proof for Janet's mnemonic formula.

2.2 Steinmetz's Symbolic Method: A Disguised Geometric Algebra

Table 2.1 reformulates Steinmetz's symbolic method in the mathematical formalism of geometric algebra.

Table 2.1: Steinmetz's Equations Reformulated in the Mathematical Formalism of Geometric Algebra

Steinmetz's Original Expressions as in His Symbolic Method	Steinmetz's Expressions Reformulated in Geometric Algebra	Comments
$\dot{E} = e^1 + je^{11}$ Voltage as a complex number. Imaginary unit squares to minus one. $j^2 = -1$ Voltage is represented in the Argand plane of complex numbers $C(1, j)$	$\bar{E} = e^1 \bar{e}_1 + e^{11} \bar{e}_2$ Voltage as a grade-1 multivector Unit vectors square to plus one. $(\lvert \bar{e}_1 \rvert)^2 = (\lvert \bar{e}_2 \rvert)^2 = +1$	Voltage as a complex number is represented in an *anisotropic* complex plane $(1, j)$. Voltage as a grade-1 multivector is represented in an *isotropic* plane of real numbers $\Re^2(\bar{e}_1, \bar{e}_2)$
$\dot{I} = i^1 + ji^{11}$ Current as a complex number. Current is represented in the Argand plane of complex numbers $C(1, j)$	$\bar{I} = i^1 \bar{e}_1 + i^{11} \bar{e}_2$ Current as a grade-1 multivector. Unit vectors square to plus one. $(\lvert \bar{e}_1 \rvert)^2 = (\lvert \bar{e}_1 \rvert)^2 = +1$	Current as a complex number is represented in an *anisotropic* complex plane $(1, j)$. Current as a grade-1 multivector is represented in an *isotropic* plane of real numbers $\Re^2(\bar{e}_1, \bar{e}_2)$
$[P] = [\dot{E}\dot{I}] \neq [\dot{I}\dot{E}]$ $[P] = [P]^1 + j[P]^j = (e^1 i^1 + e^{11} i^{11}) + j(e^1 i^{11} - e^{11} i^1)$ Steinmetz defines power $[P]$ as a double-frequency vector product of voltage and	$S = \bar{E}\bar{I} \neq \bar{I}\bar{E}$ $S = \bar{E}\bar{I} = \bar{E} \cdot \bar{I} + \bar{E} \wedge \bar{I}$ $S = P + QJ$ $P = \bar{E} \cdot \bar{I}$ as the inner product of voltage and current vectors is a scalar.	Steinmetz's expression for electrical power contains contradicting elements: • Power is considered as a vector in Gibbs' standard vector algebra • Power is expressed as a complex number in complex algebra

(Contd.)

Table 2.1: *(Contd.)*

Steinmetz's Original Expressions as in His Symbolic Method	Steinmetz's Expressions Reformulated in Geometric Algebra	Comments
current vectors. Steinmetz also represents it as a complex number. Steinmetz's double representation is mathematically ambiguous.	$P = e^1 i^1 + e^{11} i^{11}$ $QJ = \bar{E} \wedge \bar{I}$ as the outer (wedge) product of voltage and current vectors is a bivector. $QJ = (e^1 i^{11} - e^{11} i^1) e_{12}$ $J = e_{12}$ is a pseudo scalar and squares to minus one. In Geometric Algebra we define apparent power as a geometric product of voltage and current as grade-1 multivectors. Apparent power is therefore a *linear combination of a scalar (active power) and a bivector (reactive power) and it could be (geometrically) interpreted as a spinor.*	• Complex numbers are not vectors; vectors are not complex numbers • Steinmetz's expression for power is isomorphic with the canonical expression for instantaneous power: $P = \bar{E}\bar{I} \equiv p(t) = v(t)i(t)$ • The author's expression for apparent power, as a geometric product of voltage and current as grade-1 multivectors, is also *isomorphic with the canonical expression for instantaneous power* $P(t) = v(t)i(t)$ • The well-known expression for apparent power introduced by Guilbert, Janet, Breisig, and La Cour: $S = \dot{V}I^*$ is not isomorphic with the canonical expression $p(t) = v(t)i(t).$ The operation of conjugation is not justified. It is a mnemonic rule.

The following calculations contain the mathematical proof that Steinmetz's symbolic method is based on geometric algebra:

Calculation of the Inner Product: Active Power

$$\bar{E}\cdot\bar{I} = \frac{1}{2}[\bar{E}\bar{I} + \bar{I}\bar{E}] = \frac{1}{2}[(e^1e_1 + e^{11}e_2)(i^1e_1 + i^{11}e_2) + (i^1e_1 + i^{11}e_2)(e^1e_1 + e^{11}e_2)]$$

$$= \frac{1}{2}[(e^1e_1i^1e_1 + e^{11}e_2i^{11}e_2 + e^1e_1i^{11}e_2 + e^{11}e_2i^1e_1)$$
$$+ (i^1e_1e^1e_1 + i^{11}e_2e^{11}e_2 + i^1e_1e^{11}e_2 + i^{11}e_2e^1e_1)]$$

$$= \frac{1}{2}[e^1i^1(e_1)^2 + e^{11}i^{11}(e_2)^2 + e^1i^{11}e_1e_2 + e^{11}i^1e_2e_1$$
$$+ i^1e^1(e_1)^2 + i^{11}e^{11}(e_2)^2 + i^1e^{11}e_1e_2 + i^{11}e^1e_2e_1]$$

$$= \frac{1}{2}[e^1i^1 + e^{11}i^{11} + i^1e^1 + i^{11}e^{11}]$$

$$= \frac{1}{2}[2e^1i^1 + 2e^{11}i^{11}] = e^1i^1 + e^{11}i^{11}$$

The expression

$$(e^1i^1 + e^{11}i^{11})$$

in geometric algebra is *identical* to Steinmetz's expression for active power [P][1].

The expression for active power (inner product of grade-1 multivector voltage and grade-1 multivector current) represents a geometric invariant.

Calculation of the Outer (Wedge) Product: Reactive Power

$$\bar{E}\wedge\bar{I} = \frac{1}{2}[\bar{E}\bar{I} - \bar{I}\bar{E}] = \frac{1}{2}[(e^1e_1 + e^{11}e_2)(i^1e_1 + i^{11}e_2) - (i^1e_1 + i^{11}e_2)(e^1e_1 + e^{11}e_2)]$$

$$= \frac{1}{2}[(e^1e_1i^1e_1 + e^{11}e_2i^{11}e_2 + e^1e_1i^{11}e_2 + e^{11}e_2i^1e_1)$$
$$- (i^1e_1e^1e_1 + i^{11}e_2e^{11}e_2 + i^1e_1e^{11}e_2 + i^{11}e_2e^1e_1)]$$

$$= \frac{1}{2}[(e^1i^1(e_1)^2 + e^{11}i^{11}(e_2)^2 + e^1i^{11}e_1e_2 + e^{11}i^1e_2e_1)$$
$$- (i^1e^1(e_1)^2 + i^{11}e^{11}(e_2)^2 + i^1e^{11}e_1e_2 + i^{11}e^1e_2e_1)]$$

$$= \frac{1}{2}[(e^1i^1 + e^{11}i^{11} + e^1i^{11}e_1e_2 + e^{11}i^1e_2e_1)$$
$$- (e^1i^1 + e^{11}i^{11} - e^1i^{11}e_1e_2 + e^{11}i^1e_1e_2)]$$

$$= \frac{1}{2}[e^1i^1 + e^{11}i^{11} + e^1i^{11}e_1e_2 - e^{11}i^1e_1e_2 - e^1i^1 - e^{11}i^{11}$$
$$+ e^1i^{11}e_1e_2 - e^{11}i^1e_1e_2]$$

$$= \frac{1}{2}[2e^1i^{11}e_1e_2 - 2e^{11}i^1e_1e_2] = e^1i^{11}e_1e_2 - e^{11}i^1e_1e_2$$

$$= [e^1i^{11} - e^{11}i^1]e_1e_2 = (e^1i^{11} - e^{11}i^1)e_{12} = (e^1i^{11} - e^{11}i^1)J$$

The expression

$$(e^1i^{11} - e^{11}i^1)$$

in geometric algebra is *identical* to Steinmetz's expression for reactive power $[P]^j$.

The expression for reactive power (outer product of grade-1 multivector voltage and grade-1 multivector current) represents a geometric invariant.

In conclusion, the author demonstrates that underlying Steinmetz's method is a geometric algebra mathematical formalism. Steinmetz's symbolic method is neither complex algebra nor vector calculus. Steinmetz's method is based on the old Grassmann-Clifford geometric algebra; it is a geometric algebra in disguise.

It is painful to admit that, for such a long time, power engineers have been calculating power flows, stability, and state estimation without understanding the mathematical foundations of our algorithms. However, to quote Bertrand Russell (1929:58), "The fact that an opinion has been widely held is no evidence whatever that it is not utterly absurd; indeed in view of the silliness of the majority of mankind, a widespread belief is more likely to be foolish than sensible."

2.3 Rigorization of Janet's Expression

2.3.1 Hilbert Algebra

In a Hilbert space of complex vectors, the multiplication of two complex vectors – x, y follows the rule:

$$xy = x_1\bar{y}_1 + x_2\bar{y}_2 + ...x_n\bar{y}_n$$

This means that Janet's mnemonic rule conforms to the rule of multiplication introduced much later by Hilbert's algebra of complex vector space.

2.3.2 Geometric Reasoning

Represented in an Argand diagram, the voltage and the current coordinates are: $v(v', v'')$ and $i(i', i'')$ Active power as $vi \cos \varphi$ is also represented by the surface $v'i' + v''i''$; similarly, reactive power $vi \sin \varphi$ is also represented by the difference of surfaces $v'i'' - v''i'$. These surfaces correspond to geometric invariants representing active and reactive powers. The multiplication of complex voltage × complex current destroys the geometric invariance, whereas the multiplication of complex voltage × the conjugate of complex current preserves the geometric invariance. Janet's heuristic rule $S = \dot{V}I^*$ preserves the geometric invariance of the surfaces corresponding to active and reactive power.

The author's demonstration is based on a *geometric reasoning* method developed by Hongbo Li (2008).

2.3.3 A New Definition of Complex Power Based on an Extension of Complex Algebra

Andreescu and Andrica (2006) extend complex algebra and define two types of multiplication of complex numbers:

- A real product of complex numbers
- A complex product of complex numbers

Let us take two complex numbers, using Steinmetz's notation (Steinmetz, 1900a:179):

$$\dot{a} = a^1 + ja^{11} \equiv e^1 + je^{11}$$

$$\dot{b} = b^1 + jb^{11} \equiv i^1 + ji^{11}$$

The *real* product $\dot{a} \cdot \dot{b}$ is defined as:

$$\dot{a} \cdot \dot{b} = \frac{1}{2}(a^*\dot{b} + \dot{a}b^*)$$

Performing all the calculations, we obtain the result:

$$\dot{a} \cdot \dot{b} = a^1 b^1 + a^{11} b^{11}$$

The *complex* product $a \times b$ is defined as:

$$\dot{a} \times \dot{b} = \frac{1}{2}(a^*\dot{b} - \dot{a}b^*)$$

Performing all the calculations, we obtain the result:

$$\dot{a} \times \dot{b} = j(a^1 b^{11} - a^{11} b^1)$$

Replacing the symbols in these equations with Steinmetz's symbols, we obtain, apart from the sign before j in the imaginary part, Steinmetz's expressions for power:

$$\dot{a} \cdot \dot{b} = \dot{E} \cdot \dot{I} = e^1 i^1 + e^{11} i^{11}$$

and

$$\dot{a} \times \dot{b} = -j(e^{11} i^1 - e^1 i^{11})$$

Thus, we preserve the complex representation of the electrical magnitudes in AC circuits; at the same time, we maintain consistency between the expressions for instantaneous power and complex power:

$$p = vi$$

$$S = VI$$

In addition, the introduction of complex multiplication allows us to keep the expressions of power (active and reactive) consistent with the expression based on Grassmann's geometric product.

3. Conclusion

The actual paradigm used in power engineering for more than a century is based on a confusion: it mixes Janet's expression:

$$\dot{S} = \dot{V}I *$$

with Steinmetz's expression:

$$[P] = [P]^1 + j[P]^j$$

that is re-written, in all textbooks, as:

$$S = P + jQ$$

Underlying Steinmetz's algorithm or method is Grassmann-Clifford geometric algebra: the electrical entities (voltage and current) are grade-1 multivectors, and apparent power is interpreted as a spinor; active power is interpreted as a scalar, whereas reactive power is interpreted as momentum and represented mathematically as a bivector.

Underlying Janet's expression for complex power is a Hilbert algebra of complex vector spaces.

The author's contribution consists in mathematically rigorizing these two different approaches; however, this does not mean that he endorses them. On the contrary, from a physical point of view, they are incorrect. In a short form, the author's position is a mathematical rigorization but a physical refutation.

In fact, the author dissents from all the existing mathematical methods of representing electrical magnitudes in that he considers it false to represent fundamentally different physical observables (entities) as having the same mathematical identity.

Back to Steinmetz's symbolic method or Janet's ubiquitous expression, the question is not which of these two methods will prevail; the question is whether the two have to be buried as twin brothers in the same grave or in two different graves.

The conclusions of this chapter are based on a large body of literature on electrical circuits, power engineering, electromagnetism, electrical machines, and signal processing. The most important contributions and critical opinions on the power paradigm published between 1890 and 2020 were examined, as well as more than 100 textbooks.

Is the Poynting Theorem the Keystone of a Conceptual Bridge between Classical Electromagnetic Theory and Classical Circuit Theory?

1. Introduction

At the start of his investigation, the author assumed that a close relationship existed between classical circuit theory (CT) and classical electromagnetic theory (EMT). This assumption is widely shared by specialists in both CT and EMT and is supported by many textbooks and by a large body of scholarly publications.

This assumption was based on a mathematical isomorphism between Steinmetz's power expression and Poynting's vector expression. And indeed, in the mathematical formalism of Geometric Algebra (GA), this isomorphism can be expressed as follows:

Steinmetz's power expression is:

$$[VI] := V \cdot I + V \wedge I \equiv P + jQ$$

Poynting's theorem is expressed as:

$$\frac{E^2 + H^2}{2} + E \times H$$

Active power in CT is mathematically equivalent to energy in EMT:

$$P \equiv \frac{E^2 + H^2}{2}$$

Reactive power in CT is mathematically equivalent to momentum in EMT:

$$Q \equiv E \times H$$

In a similar way, many other scholarly publications support the idea that the power expression in complex form:

$$VI^*$$

is equivalent to the expression of the complex Poynting vector:

$$E \times H^*$$

2. Theoretical Debates on the Relevance of Poynting Theorem for Circuit Theory

The question of whether the Poynting vector is the keystone of a bridge between circuit theory and electromagnetic theory has long been debated in the literature. This chapter presents major contributions to that debate by both the proponents of the Poynting vector as keystone of that bridge and opponents of that view.

2.1 Proponents of the Poynting Theorem as Keystone of a Bridge between Classical Electromagnetic and Circuit Theories

Richard B. Adler, Lan Jen Chu, and Robert M. Fano, introducing the first chapter of their book *Electromagnetic Energy Transmission and Radiation* (Adler *et al.*, 1960:1), state: "It is helpful...to be able to express both lumped-circuit and field behavior in a similar language. The concept of energy ...[is] important for achieving this goal." In their book *Electromagnetic Fields, Energy, and Forces* (Fano *et al.*, 1968), these authors state that the formulae:

$$P = \Sigma I_k V_k$$

and

$$P = \oint (E \times H) \cdot nda$$

are expressions for power that can be reconciled (p. 294). They add, "The representation of field vectors in the sinusoidal steady state is an extension of the representation of voltages and currents used in circuit theory," i.e., $\dot{V}$ and $\dot{I} \Leftrightarrow \dot{E}$ and $\dot{H}$ (p. 317), and on p. 324-35, that "...the complex Poynting's vector can be identified with the complex power output to the system as defined in circuit theory:

$$-\oint S \cdot nda = P + jQ = \frac{1}{2} VI^*$$

J.M. Aller, A. Bueno, and M.E. Jimenez (1999:54-56) state, "In a three-phase transmission line...the instantaneous active power $p(t)$ corresponds to the longitudinal component of the Poynting vector, and the reactive power $q(t)$ is related to the tangential or rotational component."

M.E. Balci, M.H. Hocaoglu, and S. Aksoy (2006) state that instantaneous power is 'exactly' derived from the flux of the Poynting vector.

R. Becker (1944:195) equates the real part of the complex Poynting vector with the Joule losses and states that the imaginary part of the complex Poynting vector is identical with Slepian's expression, i.e.

$$2iw(U_{mg} - U_{el})$$

in which $U_{mg} \equiv$ energy of the magnetic field, and $U_{el} \equiv$ energy of the electric field.

Z. Cakareski and A.E. Emanuel (1999, 2001) consider that the Poynting vector interprets the physical process of power transmission geometrically.

N. Calamaro, Y. Beck, and D. Shmilovicz (2015) review the Poynting vector theorem and consider it relevant to the concept of power in circuit theory.

I. Campos and J.L. Jimenez (1992) state that the Poynting vector is an expression of energy-momentum conservation; however, they consider that the Poynting vector theory should be re-examined in relation to cases where electrical charges are present. They stress the importance of understanding the relationship between fields and matter, and they show that electromagnetic fields and charges do not constitute a closed system.

L.S. Czarnecki (2003) considers the Poynting vector relevant to power theory for balanced AC circuits, but not to power theory for unbalanced AC circuits. He and A.E. Emmanuel carried on a long scientific debate on the significance of the Poynting theorem for power theory.

F. de Leon and J. Cohen (2008, 2010), as well as A.E. Emmanuel, are some of the strongest supporters of the viewpoint that the Poynting vector is essential for power theory in electrical circuits.

F. Emde (1902, 1923) distinguishes between two hypotheses concerning power transport in electrical circuits: the first, that power is transferred *through* the conductor, and the second, that power is transferred *around* the conductor. He links the mathematical expression of the Poynting theorem with the mathematical expression of reactive power.

J.A.B. Faria (2013:367) equates the complex power expression VI^* with the complex expression of the Poynting vector, using the common formula $E \times H^*$. In his book, *Electromagnetic Foundations in Electrical Engineering* (2008), he states (p. 259), "...for time-harmonic regimes, the active power is to be physically identified with Joule losses averaged over time." He considers that "...the complex Poynting theorem... is an absolutely general theorem" (p. 260) as shown in the formula:

$p = \int_{S_T} S \cdot nds = ui.$ (p. 315). "The Poynting vector is the carrier of electromagnetic energy. Electromagnetic energy is not carried by wires. Wires are simply used to guide the electromagnetic waves. Apart from wire losses, the energy flow is essentially external to the wires" (p. 312).

J.A. Ferreira (1988) supports the thesis that the Poynting theorem is relevant for power theory. Ferrero *et al.* (2000, 2001) state that the Poynting theorem represents the bridge between electromagnetism and circuit theories.

A. Föppl (1894) questions the assumption that the energy flows from the external electromagnetic field into the electrical conductors.

W.S. Franklin (1901, 1903, 1912) refers to J.J. Thomson's critical view on the Poynting vector (i.e., that it is not uniquely defined). Franklin contradicts the ideas that voltage and current are physically equivalent to electric field and magnetic field intensities and that energy flows perpendicular to the surface of the conductor.

I. Galili and D. Kaplan (1996) underline the unity of electric and magnetic fields. They represent two sides of the same entity (electromagnetic phenomenon) that could co-exist in different proportions, depending on the observer's frame of reference. They are facets of the same object observed from different perspectives.

M. Guarnieri (2011) introduces the mathematical equivalence between voltage, current, and electric and magnetic fields as:

$$E = - \nabla V$$

$$\nabla \times V = J$$

J.D. Kraus (1991:569) equates voltage and current of the circuit with electric and magnetic fields.

$$p = VI = EHA$$

These equations are based on the following equalities:

$$V = \int E \cdot dL$$

$$I = \oint H \cdot dL$$

R. Loudon, L. Allen, and D.F. Nelson (1997:1071) state: "For the electromagnetic momentum in material media, it is necessary to take account of contributions from both the electromagnetic field and the dielectric medium."

C.G. Montgomery, R.H. Dicke, and E.M. Purcell (1948) consider the circuit theory as a part of electromagnetism; the Poynting vector is the connecting link between terminal quantities (voltages, currents) and field quantities (electrical field E, magnetic field H).

D.F. Nelson (1996:4713) concludes, "... $E \times H$ cannot be assumed to be the energy propagation vector and, in fact, is not." He adds, "...the Poynting theorem does not apply to non-linear interactions" and asserts that questions regarding energy propagation in material media "can be definitively answered only when the matter is treated on as fundamental a basis as the electromagnetic fields are..."

S. Ramo, J.R. Whinnery, and T.V. Duzer (2013:139) state that power expression in electrical circuits is given in terms of voltage and current, while the energy expression in electromagnetic fields is given by the Poynting expression in terms of electric and magnetic field intensities.

P. Russer (2003) considers that the imaginary part of the complex Poynting vector (expression) represents the reactive power that is radiated.

S.A. Schelkunoff (1948) stresses the points of contact between circuit theory and electromagnetic theory. He defines field theory as focused on electromagnetic state as a function of space, whereas circuit theory is focused on electromagnetic state as a function of time. Both field theories and circuit theories are mathematical theories, which should not be confused with circuits or fields as physical concepts. Schelkunoff supports the idea of unifying circuit theory and electromagnetic field theory.

K. Simony (1956:31) states:

$$P = UI = EHA$$

"*UI...denselben Wert ergibt als wenn wir die Leistung mit Hilfe des Poyntingschem Vektor ermittelt hätten*" [*UI* gives the same value that we would have obtained by using the Poynting vector – author's translation].

C.G. Someda (2006:57) states that "...the complex Poynting vector is

$$P = \frac{E \times H^*}{2}$$

not related to the time-domain Poynting vector by the Steinmetz method."

J.W. Simmons and M.J. Guttman (1970) state that "plane waves cannot carry angular momentum parallel to the direction of propagation" and "perfect plane waves do not exist."

M. Stone (2000) notes the importance of distinguishing between energy and momentum of electromagnetic fields and pseudo-energy and pseudo-momentum related to waves moving through a medium. He identifies Minkowski's expression with the pseudo-momentum of waves in matter (e.g., moving fluids).

J.A. Stratton (1941:133) was critical about the Poynting vector in his now classic book, *Electromagnetic Theory*: "...the validity of Poynting's theorem is unimpeachable. Its physical interpretation, however, is open to some criticism."

P.E. Sutherland (2007) states, on the basis of Poynting's complex theorem, that there are two types of reactive power.

G. Todeschini *et al.* (2007) and A. Țugulea (2002) support the thesis that the Poynting vector is relevant to power theory.

The literature investigated above supports the following tenets of the existing paradigm:

1. The Poynting theorem in classical electromagnetic theory is highly relevant for the concept of electrical power in classical circuit theory. The expression $E \times H$ signifies that: a) energy is stored in an electromagnetic field, 2) energy and momentum are 'transported' by the electromagnetic field, 3) the electric and magnetic fields are conceived as transverse infinite plane waves, 4) the direction of the Poynting vector indicates an outward flow of energy from the electromagnetic field and an inward flow of energy into the conductors of the electrical network. The electrical network is merely a huge antenna that guides the electromagnetic waves carrying the energy flow.

2. The Poynting theorem in classical electromagnetic theory represents the *keystone* of the conceptual and mathematical bridge between circuit and electromagnetic theories. Circuit theory is merely an approximation of classical electromagnetic theory; it is a sub-theory of electromagnetic theory.

3. The Poynting theorem is a *cornerstone* of classical electromagnetic theory. The Poynting theorem derives mathematically from Maxwell's equations. Therefore, as Stratton states, it is unimpeachable and supports the physically 'crazy' (Feynman, 1961) idea that energy flows from generators to the loads through space and not along the transmission lines.

The author refutes the existing paradigm and questions the soundness of the Poynting theorem and its relevance to the process of energy transfer in conductors carrying currents. Because the Poynting theorem is based on classical Maxwellian electromagnetic theory, let us ask the question: What is Maxwell's theory?

Asked the same question, Heinrich Hertz answered, "*Die Maxwellsche Theorie ist das System der Maxwellschen Gleichungen*" ('Maxwell's theory is Maxwell's system of equations') (Fölsing, 1997:371). This may be a witty aphorism, but it is also a self-referential tautology or, in plain English, a logical cul-de-sac. Moreover, a scientific theory is more than a mathematical theory; as we know, Maxwell's equations are an axiomatically defined set of equations. It is true that physics talks in the language of mathematics, but not every mathematical utterance represents a law of physics. Hertz's witticism is also misleading: in Maxwell's time, his theory was already represented in more than 20 different mathematical formalisms. J.W. Arthur (2008, 2009, 2011, 2013) gives 24 different versions of Maxwell's equations! Hertz refers to the four equations known as the Heaviside-Hertz revised version of Maxwell's equations, written in the mathematical formalism of vector calculus, in which the important physical concept of magnetic vector potential is omitted ('killed' by Heaviside).

Nowadays, when we speak about classical electromagnetic theory, we also include Lorentz's force equation, which introduces the atomistic concept in the purely 'field'-oriented Faraday-Maxwell electromagnetic theory. More than 150 years after that theory was introduced, we cannot speak anymore about one electromagnetic theory. We have textbooks on quantum electrodynamics, relativistic electrodynamics, and topological electromagnetic theories and perpetual new research and developments in the field of electromagnetism, based on relativity and quantum mechanics, question the validity of the classical Maxwell-Lorentz electromagnetic theory. In a variation of Hertz's dictum, one could say that nothing has survived from Maxwell's original theory except four equations.

For this reason, the author felt necessary to investigate the literature that takes a critical approach to Maxwellian electromagnetic theory.

2.2 Opponents' View: The Poynting Theorem is not the Keystone of a Bridge between Classical Electromagnetic and Circuit Theories

Classical electromagnetic theory considers the conductor as neutral matter. However, A.K.T. Assis and associates (Assis, 1997; Assis and Hernandes, 2007; Assis and La Mesa, 2001; Assis and Torres Silva, 2000; Assis *et al.*, 1997; Bueno & Assis, 2001; Hernandes and Assis, 2003) stated that a conductor carrying a steady current is not neutral because there are electric charges on its surface, which are unequally distributed. The strength of these charges is proportional to the strength of the electric field of the source. The charges create an electric field outside the conductor. A conductor carrying a steady state current possesses both a longitudinal electric field (parallel to the axis of the conductor) and an azimuthal magnetic field (perpendicular to the axis of the conductor). A steady current possesses an electromagnetic field that acts inside and outside the conductor.

One can conclude that the electric charges distributed along the conductor's surface act as a coating shield that will deflect the Poynting vector. Only in places where the density of the charges is null can the Poynting vector be perpendicular to the conductor's surface.

Assis and his associates refute the idea of Clausius (also supported by Feynman, Purcell and many other scientists) that a conductor carrying a steady current is electrically neutral. In their view, conductors are not merely passive antennae submitted to electromagnetic force exerted by external media. They agree with Weber (1846, 1872) and Gluckman (1999) that the conductor is also an active element that exerts an electromagnetic force on the external media.

T.W. Barrett (1993, 2000, 2008) points to the existence of electromagnetic phenomena that cannot be explained by the classical Maxwellian electromagnetic theory. These unexplained phenomena are (1) the Aharonov-Bohm (AB) and Altschuler-Aronov-Spivak (AAS) effects, (2) topological phase effects, (3) phenomena related to bulk condensed matter (Ehrenberg's and Siday's observations), (4) the Josephson effect, (5) the Hall effect, (6) the de Haas–van Alphen effect, and (7) the Sagnac effect.

Barrett has the following criticisms of classical electromagnetic theory as modified by Heaviside and Hertz:

- Heaviside and Hertz consider scalar and vector potentials as merely mathematical artifacts and therefore, omit them from Maxwell's equations; in fact, however, electrical and magnetic potentials do possess physical significance. They represent the physical gauge fields and should be expressed mathematically as local-to-global operators.
- As modified by Heaviside-Hertz, Maxwell's theory became a linear theory characterized by simple Abelian $U(1)$ symmetry. Modern electromagnetic theory extends the classical (Maxwell-Heaviside-Hertz) electromagnetic theory towards a higher, non-Abelian $SU(2)$ symmetry.
- Maxwell's classical theory is *incomplete*; it needs to be modified in order to include particles, consider the multiple connectedness of the electromagnetic space, and re-assess the electrotonic state by taking into account the existence of an electric scalar potential (φ) and a magnetic vector potential ($\mathbf{A}$).
- The concept of medium is restricted to dielectrics, in which only a displacement current exists. Classical electromagnetic theory ignores the conducting current and the conducting elements.
- Classical electromagnetic theory does not consider electrical or magnetic sources.
- Heaviside and Poynting believed that a wire functions as a sink into which energy passes from the medium (ether) and is converted into heat; for them, wires merely guide energy, with the Poynting vector pointing at right angles to the conducting wire. Barrett contradicts this position.
- When Barrett states that Maxwell's equations need extension, he is referring to the Heaviside-Hertz interpretation.

A.F. Chalmers (1973a, b; 1975, 2013) also criticizes Maxwell's theory for a number of reasons. First, he cites Maxwell's poor understanding of the conductivity phenomenon. Maxwell perceives conductivity as a discontinuity in the medium (the electrical charge) and interprets electrical current as merely a rapid change of the displacement current. Maxwell assumes incorrectly that all electrical currents correspond to the motion of electrical charges, but this is not the case for a displacement current or an induction current. Only the conduction current is equivalent to a motion of charges. Second, Maxwell did not realize that a varying current radiates. He ignored electromagnetic radiation (later demonstrated experimentally by Hertz). Maxwell restricted his analysis of electromagnetic phenomena to source-free regions of space. Third, Maxwell was a reductionist: he reduced all electromagnetic phenomena to mechanical phenomena. Finally, Maxwell misinterpreted the displacement current as equivalent to a conduction current. However, the displacement current does not involve a motion of electrons or charges.

M. Frisch (2004, 2005, 2008, 2009, 2014) considers Maxwell's theory to be an inconsistent theory. In addition, Lorentz's added expression for force does not consider the fact that an accelerating charge experiences an additional force due to the self-field. Classical electrodynamics ignores the interaction of an electrical charge with its own field (self-interaction).

P. Graneau (1984, 1991) analyzes the different expressions for electromagnetic force. He states that the Ampère expression for force between current-carrying elements shows the existence of a longitudinal component, whereas Lorentz's expression acknowledges only a transversal component. The existence of a longitudinal component explains such phenomena as rail-gun forces. He adds that classical electromagnetic theory neglects the non-local interactions between distant particles (i.e., action at a distance).

Graneau explains the functioning of induction motors as due to non-local electromagnetic effects and to the existence of the electromagnetic vector potential. Induction motor operation cannot be explained by the action of the local field or by the Poynting theorem; the energy needed to run the motor is transported across the air gap by non-local (action-at-a-distance) forces.

The electromagnetic potential provides a connection between distant particles so that the forces experienced by one particle are also felt by the others. In general, quantum mechanics involves non-local actions (e.g., the Aharonov-Bohm phenomenon), which are not explained by the classical electromagnetic theory. Like Assis, Graneau supports the Weberian electromagnetic theory and the action at a distance (non-local interaction) promoted by Ampère-Neumann-Kirchhoff-Weber.

Henning F. Harmuth (1986a, b, c; 1989, 1991, 2001) criticizes Maxwell's theory because it does not satisfy the causality principle and consequently cannot be applied to signal theory. Maxwell's theory assumes that periodic alternating electric and magnetic waves are antecedent-free: they start from zero. However, only in theory it is possible to encounter a fully formed sinusoidal electromagnetic wave that starts from zero (at $t = 0$) without having any previous values before the starting point (for $t < 0$).

Harmuth's critique of electromagnetic theory is also valid for circuit theory. As noted by Carson (1927:1), "...circuit theory explicitly ignores the finite velocity of propagation of electromagnetic disturbances." Harmuth notes the importance of recognizing that Maxwell's equations do not yield a wrong solution, but rather a solution that is undefined. Maxwell's theory fails for signal propagation in a lossy medium (Ivrlač and Nossek, 2010).

The assumption of sinusoidal electromagnetic waves (as in circuit theory) implies a periodic sinusoidal wave within an interval

$$-\infty < t < +\infty$$

Mathematically, such a wave would have infinite energy, which is physically impossible.

The assumption of planar transverse electromagnetic waves (TEM) means a one-dimensional representation of the waves similar to the voltage-current representation in circuit theory; however, in a real lossy medium, the waves are three-dimensional.

J.A. Heras (1994, 2006, 2007, 2008a, b, 2009, 2010a, b, 2011, 2016, 2017) gives a new and different explanation of the displacement current and the induction current. They are not determined only by the local and current values of the contiguous electromagnetic field. They also have a component (or a non-local term) determined by the delayed action of the global electromagnetic potential. This position is contrary to the classical Maxwellian interpretation, which states that the displacement current is equivalent to an ordinary conduction current. Heras demonstrates that the displacement current and the induction current are fundamentally different from the conduction current. In mathematical terms, Heras demonstrates that instead of differential equations, we should use integro-differential equations, which will reflect not only the time-dependent phenomena, but also the non-local (global) electromagnetic phenomena acting with time retardation.

S.E. Hill (2010, 2011) addresses a common misunderstanding of both Faraday's law and the Maxwell-Ampère law. He says that Faraday's law is often interpreted to mean that a time-varying magnetic field or flux induces a circulating electric field, i.e., that a changing magnetic field somehow causes a change in the electric field. Similarly, the Ampère-Maxwell law is interpreted to mean that a time-varying electric field causes a change in the magnetic field. He says that both laws are misinterpreted as expressing *causality*, whereas, in fact, there is only *correlation*. The principle of causality simply asserts that a cause event precedes the effect event, and if the events are separated by space, they must also be separated in time. In fact, we are dealing with a perfect correlation between a time-varying magnetic field and a time-varying electric field; both share a common cause, i.e., a time-varying current density. Since the two variations are simultaneous, it is incorrect to suppose that one could be the cause of the other.

O.D. Jefimenko (1962, 1966a, b, 2004, 2008) criticizes Maxwell's electromagnetic theory as being an 'acausal' theory. Equations linking E and H or B and D are equations of correlation and not of causation. Jefimenko (2008) supports Heras' position with regard to reformulating Maxwell's equations to take into account the magnetic vector potential and the non-Abelian, or SU(2), symmetry.

Jefimenko also stresses the fact that classical electromagnetic theory ignores the element of retardation or time delay. He points out that electromagnetic phenomena are determined not only by synchronous and contiguous conditions, but also by past and remote events.

Jefimenko considers classical electromagnetic theory as an *unfinished* theory: no physical theory is complete until or unless it provides a clear statement or description of causal links. A causal equation unambiguously relates a quantity representing an effect to one or more quantities representing a cause or causes. The principle of causality states that present phenomena are determined by previous events.

Jefimenko states that none of Maxwell's four equations defines a causal relationship; each of these equations connects quantities that occur simultaneously, thus contradicting the theory of relativity. He says that Maxwell's differential equations should be re-written as integro-differential equations. These equations should include quantities as they existed at a time prior to the time for which the quantities representing the effect are calculated.

C. Jeffries (1992, 1994) states that while the classical electromagnetic theory is based only on fields (E and H), modern electromagnetic theory is based on potentials (φ and $\mathbf{A}$). The particle-field interaction represents energy exchange between charged particles mediated by gauge fields φ and $\mathbf{A}$. The concept of particles (electrons) was foreign to Maxwell's classical theory, which envisaged the electrical field as a continuum.

Jeffries also notes that steady-state current does not mean that the charges are moving with constant velocity: the charges inside a conductor experience acceleration. Therefore, the existing model for current transmission is inadequate.

Jeffries also considers that although the Poynting theorem is a mathematically correct expression for energy conservation, it cannot be interpreted as a physical law. Poynting's vector fails to vanish in static electromagnetic fields and therefore cannot be a correct expression for energy flux. The expression

$$u_p = \frac{1}{2}\varepsilon_0(E \cdot E + c^2 B \cdot B)$$

is not a measure of physical energy, just as $S_p = \varepsilon_0 E \times B$ is not a measure of energy flux.

G. Kaiser (2004, 2011, 2012, 2015, 2016), R. Karlsson (Kaiser and Karlsson, 2005), and D. Jeltsema (Jeltsema and Kaiser, 2016) demonstrate that the Poynting complex vector (or the Poynting expression in complex vector form) is incomplete. Jeltsema and Kaiser (2016) state that the theory of instantaneous reactive power and energy is incorrect. They note that the expression $Im(E \times H^*)$, under coordinate transformations, is not an invariant.

E.J. Konopinski (1978) demonstrates the fallacy of the generally held view that the vector potential **A** has no physical meaning in classical electromagnetism. One of his most important contributions is to show that both modern electromagnetic theory and quantum mechanics should be at the same level regarding the primacy of electric and magnetic potentials.

Konopinski contradicts Heaviside and Hertz, who considered the vector potential **A** as a mathematical artifact; they ignored Faraday and Maxwell's position regarding the physical existence of an 'electrotonic' or magnetic vector potential.

Konopinski considers that replacing quaternionic algebra with vector algebra in formulating electromagnetic equations was a step backwards because it signified the mathematical transition from a non-Abelian higher symmetry back to an Abelian lower symmetry (from a non-commutative to a commutative algebra). He also stresses the importance of gauge theory for electromagnetism.

Konopinski's analysis explains the results of the Aharonov-Bohm experiment and illustrates the physical reality of the magnetic vector potential. This position corresponds with the new developments in quantum mechanics.

According to M. Kline (1962), Maxwell assumed that electromagnetic phenomena occur in ideal conditions (a vacuum). His equations do not take into account the initial conditions and the boundary limits. He therefore also assumed that the vibrations of the ether particles are purely transversal, whereas a real elastic medium can have both transverse and longitudinal waves. Therefore, Maxwell's equations cannot explain the existence of longitudinal waves, which have been proved to exist in an elastic medium.

Maxwell's theory ignores the interaction of electromagnetic waves with matter. He considered electromagnetic waves as time harmonics and plane waves, but gave no indication of how to solve the propagation of harmonic waves taking into account curved boundary conditions and curved wave fronts. Plane waves possess infinite energy: the plane wave is a highly ideal concept. No real physical source emits or sends out plane waves.

According to D.F. Nelson (1979, 1991, 1995, 1996), charges on the surface of a conductor carrying current could disrupt the tangential H

across a surface. The vector $E \times H$ cannot be assumed to represent the energy propagation vector (and in fact, it does not). The Poynting theorem is not applicable to non-linear conditions and to non-homogeneous media. Therefore, the Poynting vector is mathematically inadequate as a representation of the energy propagation phenomena. Questions about energy propagation in material media can be definitively answered only when the matter is treated as thoroughly as the electromagnetic fields are treated. The Poynting theorem does not describe the nature of the interaction between fields and matter.

P.T. Pappas (1983) presents a simple experiment that favours Ampère's original expression for force and contradicts Lorentz's expression for force. Ampere's expression for force puts in evidence the existence of longitudinal force; there is mounting evidence (e.g. rail-gun, arc discharges) for such a force. Lorentz's expression for force and Maxwell's theory ignore the longitudinal forces that exist together with the transverse force.

W. Pietsch (2012) sees the electrodynamics of the 19th century as a case of under-determination (or double ontology) between a pure field theory and the action-at-a-distance theory. After Hertz discovered electromagnetic waves, it appeared that the field theory was the victor; however, after the discovery of electrons by Joseph John Thomson in 1897, the action-at-a-distance theory experienced a revival.

Pietsch's thesis, that there are two competing electrodynamics, means that there is a scientific underdetermination perceived as particle-field double ontology. Field theory requires the existence of a continuous medium that allows for the strictly local transfer of physical actions. Action-at-a-distance theory assumes the existence of discrete or even point-like pieces of matter. The two ontologies correspond to different mathematical frameworks; field theory uses partial differential equations, whereas action-at-a-distance theory uses algebraic equations (proportions). A good example is Coulomb's law describing the force between two particles: the force is proportional to the amount of charges and inversely proportional to the square of the distance between them.

Even Maxwell admitted that the difference between action at a distance and the field view did not arise from either party being wrong. Maxwell-Lorentz electrodynamics is a field-particle theory that accepts both fields and particles as fundamental entities. In summary, following a situation of scientific under-determination, none of the theory had been abandoned. The particle-field ontology has the advantage that it permits working with both particles and fields, depending on the context.

However, the double ontology creates conceptual problems: 1) there is no agreement on the exact expression for the force of a field acting on particles, 2) there is no explanation for the recoil force that charged particles experience when they are accelerated, 3) there are open questions related to the distribution of energy and momentum between fields and particles, 4) one is confronted with two distinct laws, one for action of a particle on the field and the other for the action of the field on the particle, and 5) the Lorentz force must be supplemented by an additional force term that accounts for radiation reaction. Pietsch (2012:144) agrees with Griffiths *et al.* (2010:391), who regarded such problems as 'the skeleton in the closet of classical electrodynamics'.

Y. Pierseaux and G. Rousseaux (2006) state that longitudinal electromagnetic waves do, in fact, exist. These authors refer to the Riemann-Lorenz theory, which is based on scalar and vector potentials, and to Poincaré's theory. Both theories support the existence of longitudinal electromagnetic waves (contradicting the classical Heaviside-Hertz electromagnetism). Their contribution is also important because it contradicts the Poynting theorem and the concept of pure transverse electromagnetic plane waves. On the basis of experiments, both these authors and Rousseau *et al.* (2008) consider it necessary to reformulate the classical electromagnetic theory with regard to the primary and fundamental magnitudes E and H, which should instead be φ and $\mathbf{A}$, the scalar and the vector electromagnetic potentials.

W.G.V. Rosser (1963, 1968, 1970, 1976) criticizes Maxwell's theory for considering that electromagnetic processes (actions) take place according to Newton's law, i.e. instantaneously. The theory of relativity limits the maximum velocity of any process to the velocity of light.

The Poynting theorem appears to rest to a considerable degree on questionable physical hypotheses. However, Stratton (1941:10) proposes to retain the Poynting-Heaviside viewpoint "until a clash with new experimental evidence shall call for its revision."

3. Empirical Measurement of the Poynting Vector

Experiments that have tried to demonstrate the existence and magnitude of the Poynting vector near electrical transmission lines were inconclusive. None of the experiments performed by H. Grabinski and F. Wiznerowicz (2010), R. Helmer, L.S. Sroubova, and P. Kropic (2014), R.G. Olsen and P.S. Wong (1992), or J.W. Stahlhut, G.T. Heydt, and T.J. Brown (2007) revealed the existence of an energy flow entering the transmission line from the external medium.

4. Conclusion

1. The conductor carrying current is not electrically neutral; charges on its surface act as a coating shield and therefore would deflect the Poynting vector. The existing paradigm assumes that the energy

vector is directed perpendicular to the surface of the conductor and that it supplies the Joule losses. The proved existence of surface charges contradicts the interpretation of the Poynting vector's direction and its role as a supply of energy to the electrical network. Stratton is correct that the mathematical derivation of the Poynting vector is unimpeachable, but the physical interpretation is faulty. Conductors are not 'passive' antennae and mere receivers of energy from the external medium.

2. Classical electromagnetism contradicts the principle of causality; it considers the phenomena as occurring instantaneously and contradicting the relativity theory. Between the electric field and the magnetic field, there is a relation of *correlation*, but not of *causation*. Changes in both the electric field and the magnetic field are caused by varying charges and currents. The electric and magnetic fields are inseparable, and how much the electromagnetic field is electric and how much is magnetic is a function of the observer's position.

3. Classical electromagnetism neglects the retardation effects; all the phenomena are related by contiguous actions. In fact, the electromagnetic space is multiply connected and local phenomena are influenced by global and remote phenomena that act with time retardation. In essence, the electromagnetic theory ignores non-local causes and the retarded propagation of phenomena.

4. Classical electromagnetism ignores the self-field of the electron. Represented as a point-like particle, the electron would have infinite energy, and the electromagnetic theory does not solve this dilemma.

5. In general, electromagnetic theory considers space as free of charges. The theory works well for linear and isotropic media, but is unsatisfactory for non-linear and conductive media.

6. Classical electromagnetic theory considers the electric and magnetic fields as the primary magnitudes, neglecting the scalar and vector potentials, whereas quantum mechanics considers electromagnetic potentials as the primary magnitudes. Consequently, there is a major discrepancy between classical electromagnetic theory and quantum electrodynamic theory.

7. From the mathematical point of view, the electromagnetic theory works in an Abelian mathematical framework. Therefore, it cannot cover new phenomena, such as the Aharonov-Bohm effect, that require a non-Abelian mathematical framework.

8. Classical electromagnetic theory is mathematically undefined; it has more unknowns than equations. As a result, there are an infinite number of mathematically possible and correct solutions. To find a physical solution, additional boundary conditions are required.

9. The displacement current is incorrectly interpreted; it is not a current similar to the conduction current. The difference is that a normal

conduction current is caused by the contiguous and local phenomena, whereas the displacement current is also defined by remote (global) phenomena with retarded action. The same observation is also valid for the Faraday induction current.

10. Classical electromagnetic theory considers charges to be spatial discontinuities of the medium; the theory negates the objective existence of charges.

11. Phenomena in the conducting medium and in conductors carrying currents are not properly understood within classical electromagnetic theory.

12. Classical electromagnetic theory represents electric and magnetic fields mathematically as waves that are plane, mutually orthogonal, and sinusoidal. By ignoring the spatial boundary conditions and the fact that such waves would have infinite energy, the theory defies the physical reality.

The author considers that classical electromagnetic theory is an unfinished and incomplete theory; the Poynting theorem follows from this theory.

The existence of 24 versions of Maxwell's equations (Arthur, 2013) and 729 variations of Poynting's expression (McDonald, 2019) support the conclusion that the Poynting theorem does not represent the keystone of a conceptual bridge between electromagnetic and circuit theories. Any attempt to build such a bridge between classical circuit theory and classical electromagnetic theory (with respect to the concept of electrical power) is futile. It is like searching for an elusive *fata morgana*.

CHAPTER

4

Electromagnetic Power

1. Introduction

The author introduces a new power theory: the theory of electromagnetic power. The term 'electric power', used in classical power theory, is too restrictive. Power in circuits is neither wholly electric, nor wholly magnetic; it is related to electricity as well as to magnetism. Power in circuits is an electromagnetic phenomenon; therefore, the author replaces the term 'electric power' with the term *electromagnetic power*.

The ontic part of the theory interprets the concept of electromagnetic power in a way that is similar to the way that quantum electrodynamics interprets the concept of electromagnetic energy-momentum. This interpretation marks a leap from the classical power theory, based on classical mechanics, to a theory of electromagnetic power based on quantum mechanics and relativity.

The epistemic part of the theory uses expressions for electromagnetic power that reflect recent advances in mathematics.

2. Ontology of the New Electromagnetic Power Theory

Electromagnetic power is interpreted as the *density* in space and time (or *spacetime density*) of electromagnetic energy. Electromagnetic energy, from a relativistic point of view, resides in the four-dimensional pseudo-Euclidean Minkowski spacetime manifold.

In classical power theory (Atabekov, 1965), the instantaneous power is considered as a time derivative of energy:

$$P = \frac{dW}{dt}$$

The above mathematical expression means that between power and energy there is only a temporal relationship; however, in modern electromagnetic theory, energy has spatial as well as temporal attributes. Therefore, it is rational to interpret electromagnetic power as having both temporal and spatial attributes. The author interprets electromagnetic power as a spacetime derivative of electromagnetic energy.

The new theory interprets electromagnetic power as the *density* of electromagnetic energy in spacetime; the electromagnetic power is limited in space and in time. It is conceived as the energy density at a specific point in space and at a specific moment in time.

Electromagnetic power is *localized in space*, in a space interval:

$$(x \pm \Delta x \to 0)$$

Electromagnetic power is *constrained in time*, in a time interval:

$$(t \pm \Delta t \to 0).$$

Electromagnetic power is perceived in space at the terminals (nodes) of the electrical network. Electromagnetic power is perceived in time, at a specific moment, as *instantaneous* power.

3. Epistemology of the New Electromagnetic Power Theory

In the new theory, electromagnetic power has both spatial and temporal attributes; it represents mathematically the full spacetime derivative of the mathematical function representing electromagnetic energy.

In the theory of generalized functions, the electromagnetic power as a *spacetime density* of energy can be represented as a *functional* – a function of functions. The mathematical branch dealing with generalized functions and functionals is the theory of distributions (Berg, 1965; Lützen, 1982; Yosida, 1984), and it represents an interesting avenue for research regarding the epistemology of electromagnetic power. However, this direction of research is outside the scope of this monograph.

The present author uses the mathematical formalism of geometric algebra, in which electromagnetic power is equivalent to the Faraday electromagnetic *force* in modern electromagnetic theory. Physically, the author interprets the *apparent power* as an electromagnetic force; mathematically, he represents it as a multivector: a linear combination of a *scalar* and a *trivector*. The scalar represents the density of potential energy corresponding to the term *active power*. The trivector represents the density of kinetic energy (or the density of the electromagnetic momentum) and corresponds to the term *reactive power*.

4. The Main Characteristics of the New Electromagnetic Power Theory

The theory is applicable to DC and AC (sinusoidal and non-sinusoidal linear) circuits. It bridges the gap between circuits and fields, giving a common understanding of the concept of power and energy for both circuit theory and electromagnetism.

It re-interprets the familiar concepts of the existing paradigm physically and mathematically:

- *Apparent* power as density in spacetime of the Faraday electromagnetic force.
- *Active* power as density in spacetime of the electromagnetic potential energy, expressed as a *scalar*.
- *Reactive* power as *density* in spacetime of the electromagnetic kinetic energy, expressed as a trivector.

The present author's expression for electromagnetic power as a linear combination of a scalar and a trivector is similar to the tensorial expression for energy-momentum in classical electromagnetic theory. The electromagnetic power theory dissents from the classical power theory (paradigm) in stating that:

- Apparent power has a *physical interpretation* and is not merely a design quantity.
- Reactive power is a *density of electromagnetic momentum;* electromagnetic momentum corresponds to kinetic energy.
- Reactive power is not a concept related to AC circuits only. Reactive power as *momentum* exists in both AC and DC circuits. The difference is that in DC circuits there is only a linear momentum, whereas in AC circuits there are two momenta: a linear momentum and a rotating momentum. The combination of these two momenta results in a *helicoidal trajectory of energy* around or wrapping the conductors. The energy transfer in power lines follows a helicoidal trajectory.

This new theory defines power as the geometric product of a *voltampère* multivector and its reverse. Compared with the Blakesley-Ferraris expression:

$$p = vi$$

the author's axiomatic definition is:

$$p = (v + Ii)\,(v + Ii)^{\dagger}$$

Steinmetz considered voltage and current to be epiphenomena and represented them with identical mathematical symbols (either complex numbers or vectors). This author adopts Steinmetz's position that voltage and current are epiphenomena, but represents them with different mathematical symbols: a (polar) vector for voltage and a bivector for current. This mathematical representation is similar to the representation of electric and magnetic fields in quantum electrodynamics, where the intensity of the electric field is depicted as a vector, and that of the magnetic field, as a bivector.

The author contradicts the classical assumption of voltage and current *separability*; voltage and current are Janus' faces of electromagnetic power. Moreover, on the basis of the relativity theory, what is an electric field for one observer is a magnetic field for another observer, and similarly for circuits, what is voltage for one observer is current for another. The author introduces a new multivector that is a *linear combination* of voltage and current. Voltage is expressed as a 1-grade multivector, whereas current is expressed as a 2-grade multivector (or bivector). The new multivector is called *voltampère*. The expression for electromagnetic power (shown above) is the geometric product of this multivector and its reverse:

The expression:

$$v + Ii$$

is similar to the expression of the electromagnetic field:

$$E + IB$$

5. Geometric Algebra in Electrical Engineering and Power Theory

5.1 Pre-history of Geometric Algebra in Mathematics

In addition to the publications cited below, the author consulted the following works for general historical background: Pawlikowski (1967), Gull (1993), Moore (1995), and Gürlebeck (2008).

In 1832, when Hermann Grassmann conceived his non-commutative vectorial system, he was unaware of Möbius' work, *Barycentrischer Calcul* (Barycentric Calculus), published in 1827. Möbius was one of the first mathematicians to introduce the concept of a directed line that contains the idea of non-commutativity. In his book, Möbius defined a line segment from a point A to a point B as AB and stated that $AB = - BA$. Grassmann, in his *Ausdehnungslehre* (Theory of Extension), published in 1844, rediscovered the barycentric mathematical system, which represents a linear combination of points, and reinterpreted it as a linear combination of vectors. Möbius' mathematical system provided an algebra of points, while Grassmann's mathematical system provided an algebra of vectors (Dorier, 1995). Hermann Grassmann was also unaware of another development: Gauss's geometrical interpretation of the addition of complex numbers, which is similar to the addition of vectors.

During the period 1839-1844, Grassmann fully developed his mathematical system of n-dimensional spatial analysis based on the concept of vectors. He implemented his father's idea of a geometrical product (Justus Günther Grassmann, 1824) by stating that the geometrical product of vectors "represents the surface content of the parallelogram determined by these vectors" (Crowe, 1985:61). It is significant to mention that Grassmann's geometrical product of two vectors (outer product) represents an oriented area (bivector), which is fundamentally different from the vector or cross product in Heaviside-Gibbs vector calculus. In vector calculus, the geometric (cross, outer, skew) product of two polar vectors results in an axial vector perpendicular to the area defined by the two polar vectors.

It is important to stress that although the term *vector* used in the above paragraphs was introduced in 1845 by William Rowan Hamilton, the concept of a vector as a directed line segment had been known since 1827 from the barycentric calculus of Möbius. The term had also appeared in 1776 in Diderot's *Encyclopédie*; it is derived from *vectus*, the past participle of the Latin verb *vehere*, 'to carry'. In 1843, Hamilton had invented his vectorial systems: the quaternions. He introduced the operation of quaternion multiplication, which gives two parts: the scalar product (denoted $S\alpha\beta$) and the vector part (denoted $V\alpha\beta$). It is important to stress that Hamilton's concept of a *vector* is completely different from the concept of a *vector* or its interpretation in Grassmann's vectorial system or from that in Heaviside-Gibbs vector calculus. The different interpretations of the term *vector* have created a lot of confusion and triggered an unnecessarily long and bitter debate between quaternionists and vectorists (Altmann, 1986, 1989; Bork, 1966, 1967).

In 1852, for the first time, Hamilton read Grassmann's book, *Ausdehnungslehre* (1844) and he highly praised Grassmann as a great genius. Hamilton mistakenly identifies Grassmann's inner product with his scalar part of the quaternionic product and Grassmann's outer product with his vector part of the quaternionic product.

Grassmann, a high-school teacher, did not possess Hamilton's scientific fame and authority. Furthermore, his book and publications were difficult to read as his style was heavy and overloaded with philosophical ideas (influenced by Schleiermacher's *Dialektik*). For these reasons, Grassmann's ground-breaking ideas related to an *n*-dimensional vectorial system found little resonance or acceptance (Hestenes, 2011).

In physics, Grassmann-Clifford algebra was rediscovered in the matrix algebra of Pauli and Dirac, where it plays an essential role in quantum mechanics (Petsche, 2011). In power engineering, Steinmetz unwittingly rediscovered geometric algebra (Petroianu, 2014; Sangston, 2016).

One of the few promoters of Grassmann's ideas was Hankel, who published an influential book, *Vorlesungen über die komplexen Zahlen und ihre Funktionen* ('Lectures on Complex Numbers and Their Functions'), in 1867. Hankel praised Grassmann and stressed the fundamental importance of his theory of extension (Hankel, 1867).

Grassmann's theory of extension also made a big impact on Felix Klein and on his Erlanger Program (1872). From his side, although critical with respect to quaternions, Josiah Willard Gibbs, in his seminal work, *On Multiple Algebra* (1886), is full of praise for Grassmann's work (Gibbs, 1886).

William Kingdon Clifford (1845-1879) played an important role in the development of Grassmann's vectorial system. Familiar with both Hamilton's quaternions and Grassmann's theory of extension, Clifford introduced the *inner* product of vectors one year before his death (Clifford, 1878).

Clifford referred to his algebra, which is applicable in n-dimensional spaces, as geometric algebra; however, in some publications, this algebra is still defined as Clifford algebra. The *inner* product is interpreted as a *scalar* representing a *volume*; the *outer* product is interpreted as a *directed area*. Both products are *distributive*, and their linear combination – the *geometric product* – is *invertible*.

Two years after Clifford, Rudolf Lipschitz independently discovered the same algebra. Without diminishing the important role played by Hestenes in reviving the Grassmann-Clifford geometrical system, it would be unfair to ignore Lipschitz's contribution and to define this vectorial system as just Hestenes' *Geometric Algebra*. The important role played by his paper, *Principes d'un calcul algébrique qui comme espèces particulières des quantités imaginaires et des quaternions* (Lipschitz, 1880) is recalled by Marcel Riesz (1993).

On the other hand, Hestenes warns: "Do not confuse geometric algebra (GA) with Clifford Algebra (CA)!"(Hestenes, 2017:352).

Dorier (1995) defines Grassmann-Clifford geometric algebra as a dialectic contradiction between algebra and geometry.

At the end of 1880, there were two competing vectorial systems: Hamilton's quadruple algebra called quaternions and Grassmann-Clifford mathematical system. In addition to these systems, J. Willard Gibbs mentions the contribution of the Frenchman, Saint-Venant. In his presentation on *'Multiple Algebra'* in 1886, Gibbs states: "It is a striking fact in the history of the subject that the short period of less than two years was marked by the appearance of well-developed and valuable systems of multiple algebra by British (Hamilton, 1843), German (Grassmann, 1844), and French (Saint-Venant, 1845) authors, working apparently entirely independent of one another" (Stephenson, 1965).

Independently and quasi simultaneously, Oliver Heaviside and Willard Gibbs developed the 'modern' *vector calculus*. Heaviside was motivated by his desire to simplify Maxwell's equations; Gibbs (1891) considered that the 'union' of the two products, in Hamilton's quaternionic algebra, does not permit a good geometric insight. For these reasons, Heaviside and Gibbs 'separated' the geometric product into two parts: the scalar (dot, inner) product and the vector (outer, skew, cross) product. The scalar part of the quaternions had a negative sign, whereas the dot product of the new vectors had a positive sign (O'Neil, 1986).

Unlike the dot product, the cross product is not commutative, but anti-commutative. The cross product does not admit generalization in a higher order space; the cross product is valid only in three-dimensional and seven-dimensional space (Craig, 1951; O'Neil, 1986).

Hertz strongly supported the reformulation of Maxwell's equations in the mathematical formalism of vector calculus, and the German school of electromagnetism, represented by Föppl and Abraham, adopted vector calculus and influenced Steinmetz in his vectorial treatment of power in AC circuits.

Considered as 'heretics' with respect to the orthodox view of quaternions, Gibbs and Heaviside were severely criticized by Hamilton followers, such as P.G. Tait (1886) and C.G. Knott (1893). Added to the heretics was also Alexander Macfarlane with his 'hyperbolic quaternion'. Macfarlane has the merit of exposing the conceptual mixture in quaternionic algebra which comprises the use of the same symbol to represent both a quadrantal versor and a unitary vector. In fact, the 'vectors' *I*, *J*, and *K* in quaternionic algebra are *axial vectors or bivectors* (Macfarlane, 1893). Moreover, as Altmann (1989) points out, they represent not quadrantal (90°) rotation, but binary (180°) rotation. Altmann also indicates the important role played by Olinde Rodrigues (1840), who, independent of Hamilton, developed his theory of rotations in three-dimensional space. It is important also to mention that quaternionic algebra and vector calculus differ in terms of their chirality properties (right-handed versus left-handed) (Altmann, 1989; Gray, 1980). And so began a debate called 'the big quaternionic war' or the 'struggle for existence – *Kampf ums Dassein*' (Bork, 1966, 1967). The debate lasted for years, spanning eight scientific journals and involving many leading scientists (Wisnesky, 2004; Saá, 2007). After a heated debate, by 1894, the Gibbs-Heaviside *vector calculus* prevailed. Vector calculus became the standard mathematical tool for physicists and engineers.

It is important to stress that vector calculus does not fulfill Benjamin Peirce's (1881) definition of algebra because neither the scalar product nor the cross product lies inside the vector domain. The result of the dot product is a scalar outside the vector domain. The result of the cross product of two polar vectors is an axial vector or a bivector.

Although cross products are useful in physical applications, the cross-product operation cannot be generalized to any multi-dimensional space. Unexpectedly, as demonstrated by Eckmann (1943, 2006), the operation is possible only in three-dimensional or seven-dimensional spaces (*see* also Silagadze, 2002).

At the end of the pre-history of vectorial systems (Farouki, 2008; Perwass, 2009; Stephenson, 1966; Xambó-Descamps, 2018), we have three competing schools:

- Grassmann-Clifford-Lipschitz geometric algebra
- Hamilton quaternionic algebra
- Gibbs-Heaviside (modern) vector calculus

5.2 The History of Geometric Algebra in Electrical Engineering

E.A.F. Punga (1879-1962) can be considered as the first to promote the use of Grassmann's vectorial system in power engineering. In an article published in *Zeitschrift für Elektrotechnik*, Punga (1901) strongly recommends Grassmann's theory of extension as a mathematical tool necessary in the investigation of power phenomena in AC circuits. In support of his approach, he refers to similar opinions expressed by A.E. Blondel and H.J. Görges. Interestingly, the editor was reluctant to publish Punga's paper, commenting that the abstract mathematical content of the paper was outside the profile of a strictly engineering journal, such as *Zeitschrift für Elektrotechnik* (today, *Elektrotechnik und Maschinenbau*).

G.W. Lewis proposed to investigate electrical phenomena by using Minkowski's four-dimensional spacetime and underscoring the superiority of Grassmann's vectorial system over the Gibbs vector calculus. He considers that Minkowski's vector analysis in a four-dimensional spacetime "restores many features of the original, and much neglected, system of Grassmann" (Lewis, 1910:165).

H.W. Nichols (1917) criticized Steinmetz's symbolic method; he considered multiplication of voltage and current as complex variables to be devoid of physical meaning. Nichols proposed a new mathematical expression for electrical power that is very close to the Grassmann-Clifford geometric product of multivectors.

E. Folke Bolinder attended the private lectures of Marcel Riesz on Clifford Algebra held at the University of Maryland in 1957-1958. Together with Pertti Lounesto, Bolinder edited and published Riesz's lectures under the title, *Clifford Numbers and Spinors* (M. Riesz, 1993). Bolinder and Lounesto have the merit of making Riesz's work available to a larger audience and popularizing Clifford Algebra. Years later, Lounesto published his own important contribution to Clifford Algebra (P. Lounesto, 2001).

It is Bolinder's merit to have introduced Clifford Algebra to the electrical engineering community by rekindling interest in a geometrical approach in electrical engineering by publishing numerous research articles on microelectronics, electromagnetism, and antenna. Bolinder uses Clifford Algebra in his research and recalls the important contribution of Rudolf Lipschitz who, unaware of Clifford's work, rediscovered the same algebra (Lipschitz, 1886).

5.3 Applications of Geometric Algebra in Power Theory

Geometric algebra comprises a large number of algebraic systems, such as R - algebra of real numbers, C - algebra of complex numbers, Q - Hamiltonian algebra of quaternions, matrix algebra (Sobczyk, 2019; González Calvet, 2010), Gibbs-Heaviside vector calculus, affine geometry (Crainic and Petroianu, 2006), and projective geometry.

In addition, geometric algebra permits analysis in spacetime,, multidimensional spaces, and non-Euclidean geometries.

After publishing Marcel Riesz's lectures on Clifford Algebra, as noted above, E. Folke Bolinder became a strong promoter of geometric algebra in Electrical Engineering.

The present author has had a long-time interest in geometrization of the power concept (Petroianu, 1969; Crainic and Petroianu, 2006 a, b) and in the concept of reactive power (Fetea and Petroianu, 2000 a, b). This interest was rekindled by the book on geometric algebra by Ramon González Calvet, who states that energy losses in a transmission line are calculated as the inner product of the voltage and current 1-grade multivectors (González Calvet, 2000:19, footnote 1).

He also benefited from personal correspondence with Prof. González Calvet (Institut d'Estudis Catalans) and from the results of the Clifford project initiated by Josep Parra-Serra, professor of physics at the University of Barcelona (González Calvet, 2013; Parra-Serra, 2009).

During the last two decades (2000-2020), applications of geometric algebra in electrical engineering and electromagnetism were developed in parallel at several universities in various countries:

1. University of Adelaide (Australia): J.M. Chappell, A. Iqbal, D. Abbott, and N. Ianella
2. University of Almería (Spain): F.G. Montoya, A. Alcayde, F.M. Arrabal-Campos, and Raul Baños
3. University of Calgary (Canada): M.D. Castro-Nuñez
4. University of Delft (Netherlands): A. Chaves and D. Jeltsema (the latter is now at Hogeschool Arnhem and Nijmegen (HAN) in Arnhem (The Netherlands)
5. Northeastern University (Boston, USA): H. Lev-Ari and Tufts University (Medford, MA. USA): A.M. Stanković

6. University of Patras (Greece): A. Menti, T. Zacharias, and J. Milias-Argitis
7. National Defence Academy (Khadakwasla, India): K. Muralidhar
8. University of Seville (Spain): Juan Carlos Bravo Rodriguez, Manuel Castilla-Ibáñez, Manuel Ordoñez-Sanchez, and Juan Carlos Montaño-Asquerino
9. Tennessee Technological University (Cookville, TN, USA): Sosthenes Karugaba, and Olorunfemi Ojo,

as well as by independent researchers, such as Alexander Arsenovic and Peeter Joot.

The following section summarizes the main contributions of these researchers.

Alex Arsenovic (2017) of 810 Labs, Stanardsville, VA (USA) uses conformal geometric algebra to solve an old problem in circuit analysis – impedance matching. Conformal geometric algebra (CGA) is an extension of geometric algebra ($G_{3,0,1}$) from the 3-dimensional Euclidean space to a 5-dimensional space ($G_{4,0,1}$); conformal geometric algebra is also used in problems related to robotic vision. Arsenovic (2017, 2019) reformulates the microwave network theory in the mathematical formalism of geometric algebra.

Juan Carlos Bravo-Rodríguez

Under the guidance of Prof. Manuel Castilla-Ibáñez at the University of Seville, Juan Carlos Bravo-Rodriguez began in 2006 to investigate geometric algebra and its application in power theory. In 2008, he defended his Ph.D. thesis, entitled *Representación multivectorial de la potencia aparente en regímenes periódicos n-senoidales aplicando álgebras de Clifford* ('Multivectorial Representation of Apparent Power in *n*-sinusoidal Periodical Operating Conditions Using Clifford Algebras').

In the period from 2007 to 2018, he also produced an impressive number of articles together with Prof. Castilla, Manuel Ordoñez Sanchez, and Juan Carlos Montaño. These authors represent voltage and current as 1-grade multivectors. Therefore, apparent power, which is the geometric product of voltage and current multivectors, consists of a complex scalar plus a complex bivector. The authors generalize geometric algebra and introduce a new geometric product: the generalized complex geometric product. The distorted power is represented as a complex geometric product.

Milton Castro-Nuñez

In 2013, Castro-Nuñez defended his doctoral thesis, *The Use of Geometric Algebra in the Analysis of Non-sinusoidal Networks and Construction of a Unified*

Power Theory for Single-phase Systems – A Paradigm Shift, at the University of Calgary. From his thesis, he derived a number of publications (Castro *et al.*, 2010; 2012 a, b; 2016).

Castro-Nuñez's dissertation contains some controversial opinions:

- Castro-Nuñez accepts Steinmetz's method for circuits under sinusoidal operating conditions, but opposes it for circuits under non-sinusoidal operating conditions. The present author disagrees; he considers that Steinmetz's symbolic method for describing the concept of power in electrical circuits mathematically and interpreting it physically is incorrect under both sinusoidal and non-sinusoidal operating conditions.

- Castro-Nuñez's formula for time-space transformation represents an egregious mistake. Time and space are not dual mathematical entities, and oscillations in time are not equivalent to rotations in 2- and 3-dimensional space or reflections in $n > 3$ dimensional space.

- Castro-Nuñez equates the ith harmonic with an i-graded bivector. The frequency of the harmonic is erroneously equated with a spatial rotation, e.g., the second harmonic corresponds to a bivector (a two-fold spatial rotation), the third harmonic with a trivector (a three-fold spatial rotation), and the nth harmonic with an n-fold spatial rotation. In essence, Castro-Nuñez considers power to be different at different frequencies and gives a different geometric representation to each harmonic power. The present author strongly disagrees: nth harmonic oscillation in time is not equivalent to an n-graded multivector in space.

- Castro-Nuñez denies the equivalence between time-domain analysis and frequency-domain analysis; hence he implicitly denies the validity of Fourier analysis. He considers his finding as an 'astonishing' discovery because "the [equivalence of time domain and] frequency domain has never been challenge[d] before" Castro-Nuñez (2013:107). The present author strongly disagrees: Fourier analysis remains an unimpeachable, rigorous domain of mathematics and continues to be used by many authors (e.g., Gröchenig, 2001; Hahn and Snopek, 2017).

- Castro-Nuñez misconceives the applicability of complex algebra; he states that 'phasors' of different speed rotations cannot be added. He describes this as the 'additiveness problem'. The present author strongly disagrees: in reality, complex numbers are not rotating vectors and are not related to time oscillations or to rotations in space. Complex numbers are operators, and their multiplication is equivalent to an extension (or contraction) and a rotation in the Argand plane. Liu and Heydt (2005) in their Fig. 4, "A pictorial illustrating $v(t) = \cos(2\pi f_s t) + m \cos(2 \cdot 2\pi f_s t + \theta)$ as a resultant sum of two rotating vectors", provide a clear example in which phasors of different frequencies (represented by complex numbers) are added for a specific moment in time (a snapshot).

- Castro overemphasizes the importance of geometric algebra; he considers this mathematical formalism as a condition *sine qua non* for research in electrical circuit analysis and power theory. The present author disagrees: geometric algebra is a powerful mathematical tool, but it is not the only available instrument in the engineering toolbox. Other interesting mathematical formalisms (e.g., differential forms, functionals, and distribution theory) could also be deployed for research in power theory.

- In conclusion, the publications of Castro-Nuñez promote application of geometric algebra in electrical engineering. However, his Ph.D. thesis contains a number of misconceptions, and this author therefore rejects his claim that it represents a 'paradigm shift' in power theory.

James M. Chappell

J.M. Chappell *et al.* (2014) have provided a very good introduction to geometric algebra for electrical engineers. However, this publication reflects just one aspect of intense research activity at the University of Adelaide focusing on applications of geometric algebra to relativity and electromagnetism (Chappell *et al.*, 2010; 2012; 2016).

Dmitri Jeltsema

Prof. Dimitri Jeltsema's research activity in mathematical modeling of physical systems and in power theory (reactive power in circuits operating under non-sinusoidal conditions) is very impressive (Jeltsema and Kaiser, 2016; Jeltsema *et al.*, 2014). However, his contribution to geometric algebra and its applications in electrical circuits is limited to only two publications (Chavez-Jímenez and Jeltsema, 2011; Chavez-Jimenez *et al.*, 2011).

Peeter Joot

Peeter Joot has to be commended for his contribution to the education of electrical engineers regarding applications of geometric algebra in electromagnetism (Joot, 2009, 2015, 2017, 2018, 2019).

S. Karugaba and O. Ojo

The paper by Sosthenes Karugaba and Olorunfemi Ojo of Tennessee Technological University (Karugaba and Ojo, 2009) is the only publication known to this author that applies geometric algebra to determination of active and reactive powers in electrical machines.

Hanoch Lev-Ari and Aleksander M. Stanković

Hanoch Lev-Ari (Northeastern University) and Aleksandar M. Stanković (Tufts University) have made a major contribution to the development of power theory in circuits operating under non-sinusoidal operating conditions (e.g., the 7th-component decomposition of power) (Lev-Ari and Stanković, 2006) and to the geometric interpretation of the concept of power in a Hilbert space (Lev-Ari and Stanković, 2003) and in a Euclidean waveform space. They published only two papers applying geometric algebra to power theory (Lev-Ari and Stanković, 2009 a, b). In their geometric interpretation, voltage and current are 1-grade multivectors.

Anthoula Menti

Menti *et al.* (2007) of the University of Patras in Greece published the first paper in the power engineering literature to deal with the application of geometric algebra in electrical circuits working under non-sinusoidal conditions. A slightly modified version of this paper was presented at a conference in Lagow, Poland, in 2010 (Menti *et al.*, 2010). This paper had a large resonance and triggered wide interest among other researchers for geometric algebra applications in power theory.

Menti *et al.* interpret geometric algebra as an adequate tool for analysis of circuits operating under non-sinusoidal conditions, just as complex algebra is an adequate tool for analysis of circuits operating under sinusoidal conditions. They represent voltage and current as 1-grade multivectors and as waveforms considered to be members of the same functional space.

The present author dissents for two reasons. First, complex algebra is not an adequate mathematical tool even for analysis of electrical circuits operating under sinusoidal conditions; second, voltage and current are not members of the same functional space. A required condition of being members of the same vector space is that the vector space be closed under vector addition. However, every electrical engineer knows that you cannot add a voltage vector to a current vector. Therefore, the authors' assumption that voltage and current belong to the same functional space is incorrect. The authors do not discern the 'physical' difference between voltage and current that places them in different vectorial spaces.

F.G. Montoya et al.

In Spain, a group of researchers at the University of Almería (F.G. Montoya, A. Alcayde, F. Arrabal-Campos, and Raul Baños) followed the path opened by Castro-Nuñez (University of Calgary) and advanced the idea of using geometric algebra to represent inter-harmonics in circuits operating under non-sinusoidal conditions. (Montoya *et al.*, 2010; 2019)

The present author disagrees with several ideas contained in these publications. The first is their statement that apparent power is an artificial mathematical concept, which implies that it has no physical existence. The second is their proposal to map the continuous spectrum of inter-harmonics on to the matrix structure of geometric algebra. The third is their proposed time-space transformation, which is egregiously incorrect because time oscillations (e.g., vibrations) are not equivalent to space rotations. Rotations are possible only in two- and three-dimensional spaces. In *n*-dimensional spaces, we deal with reflections.

These authors perceive an inequality between time-domain analysis and frequency-domain analysis and give preference to frequency-domain analysis. In fact, however, the frequency domain and time domain are dual domains of analysis, and the results obtained in one domain are perfectly replicated in the other one.

Finally, Montoya *et al.* consider geometric algebra to be a theory that fills the gap between the algebra of complex numbers and other theories. In fact, however, geometric algebra is not a theory: it is merely a tool and another mathematical formalism.

Kundeti Muralidhar

Kundeti Muralidhar of the Physics Department, National Defence Academy, in Khadakwasla, India, has produced some very insightful writings (Muralidhar, 2011, 2015a, b; 2016, 2017, 2018). Although Muralidhar focused on electromagnetic theory, the idea of a complex vector (the sum of a vector and a bivector) is applicable to electrical circuits and power theory as well. The idea of representing the electrical field and magnetic field together as a complex vector (known as the Riemann-Silberstein vector) was first introduced by Silberstein (1907a, b; 1914; 1922), discussed by Bateman ([1915] 1955; [1932] 1944) and promoted later by Iwo and Zofia Bialynicka-Birula (2006, 2012, 2013) and Kaiser (2004).

5.4 Conclusions from Literature on Geometric Algebra in Power Theory

Common to these authors are the following positions:

1. They state that electrical power in circuits is the geometric product of voltage and current interpreted in geometric algebra as one-grade multivectors.
2. They accept Steinmetz's symbolic method and his expression for power in circuits operating under sinusoidal conditions, but not in circuits operating under non-sinusoidal conditions. The expression for power in circuits operating under non-sinusoidal conditions requires mathematical formalism of geometric algebra.

3. Their research is confined to power theory (and especially the theories of reactive power and reactive distortion) in electrical circuits operating under non-sinusoidal conditions. They examine power theory only in AC circuits.

4. They doubt the equivalence between frequency-domain analysis and time-domain analysis.

5. They equate oscillations in time (harmonics) with rotations in space; each time harmonic is endowed (in their approach) with a rotation in the space of geometric algebra.

6. The focus of their practical application is on reactive power compensation.

7. They consider apparent power to be a notion devoid of physical content and just a mathematical artefact.

8. They conceive the existence, in circuits operating under non-sinusoidal conditions, of ith voltage harmonics without an ith current harmonic and vice-versa. They also conceive the existence of power as the product of voltage (of one frequency) and a current (of a different frequency). They assume that a group of current harmonics could exist without corresponding voltage harmonics, and vice-versa (i.e. existence of power resulting from cross-products of voltage and current of different frequencies).

9. Their attempt to 'unify' circuit theory and electromagnetism is based on the apparent similarity between the expression of the complex Poynting vector $p = \dot{\mathbb{E}} \times \mathbb{H}^*$ and Janet's expression $\dot{S} = \dot{V}I^*$. However, this idea is misguided.

In contrast, the present author asserts the following fundamental theses:

- There is a single electromagnetic power theory covering the physical concept of power in DC and AC electrical circuits operating in any frequency range and under any operating conditions, i.e., sinusoidal, non-sinusoidal, balanced and unbalanced, linear and non-linear conditions.

- Many physical interpretations and many mathematical representations of the same concept, electromagnetic power, are possible. The author supports a pluralistic view with regard to the ontology and epistemology of power theory.

- The author's position is guided by the unified theory of forces in Nature: electromagnetic, weak, strong, and gravitational forces. In the near term, his focus is to establish the unity between circuit theory and quantum-relativistic electromagnetic theory. His goal is to develop a theory of electromagnetic power within a quantum-relativistic circuit theory and to elucidate the process of energy transmission at the mesoscopic level (i.e., at the level of quantum processes occurring in matter at the subatomic scale (within circuits and apparatus of the electrical network).

- This approach requires a transition from classical mechanics towards quantum mechanics and restraint relativity.
- We should discard both the classical (Blakesley-Ferraris) definition of instantaneous power as the product of voltage and current and the mathematical expression of electrical power based on Steinmetz's symbolic method or Janet's mnemonic formula and replace them with expressions based on geometric algebra or expressions based on differential forms, distribution theory, or functionals.
- Steinmetz's symbolic method is simply a calculation algorithm; it is a mathematically inconsistent system that mixes complex algebra with vector calculus (two incompatible mathematical formalisms). Steinmetz obtained his expression for complex power by 'sleight of hand' (assuming that for double-frequency processes, the square of the imaginary is equal to unity). Janet's heuristic expression is based on the assumption (unknown to him at that time) that voltage and current are complex vectors in Hilbert's complex vector space. The author considers that the approaches of Steinmetz and Janet have both served well for more than a century, but at present, they are obsolete and have to be replaced with a different mathematical formalism. Although the author is using geometric algebra (in this monograph), this does not preclude the use of other mathematical systems.
- The above-mentioned applications of geometric algebra are simply re-hashing the old mathematical symbolic method; instead of vectors (in vector calculus), the voltage and current are redressed as 1-grade multivectors. This is a mathematical retrofit (like installing solar panels on a 120-year-old building) instead of a change in the underlying concept.
- The author rejects the concept of 'separability' of voltage and current. They are not two independent variables; they are 'observables' of epiphenomena. There is not a causal relationship between power and voltage and current. Power is not 'produced' by voltage and current. On the contrary, voltage and current are the 'faces' of electromagnetic power; they are merely correlated with power.
- The author's new expression for electromagnetic power mimics the expression of the Faraday vector or Riemann-Silberstein complex expression for electromagnetic energy and momentum. The author considers that, as in quantum electrodynamics, electromagnetic force is expressed as the geometric product of $(E + IH) (E + IH)^\dagger$, and in a similar way, electromagnetic power is expressed as the geometric product of $(v + Ii) (v + Ii)^\dagger$. The complex vector $v + Ii$ is called *voltampère* and is a linear combination of 1-grade voltage multivector and ampère bivector pre-multiplied by the pseudo-scalar (I).

- A fundamental difference between the existent power expression and the new expression for electromagnetic power consists in the 'inseparability' of voltage from current (or vice-versa). Voltage and current form a unique physical and mathematical entity: a linear combination of a vector and a bivector. The representation of voltage as a vector expresses the non-rotational structure of the *electric* field, whereas the representation of the current as a bivector expresses the rotational structure of the *magnetic* field. Existing power theories do not consider the fundamental physical difference between voltage and current, and they assign the same mathematical identity (as vectors, or complex numbers, or sinusoidal waves) to these two different physical entities.

- This author disputes the expression of power obtained from the trigonometric representation of voltage and current as sinusoidal waves; the multiplication of two sinusoids results in a trigonometric function of double frequency. On the basis of this mathematical artefact, we obtain an active power component and a reactive power component of double frequency. This result brings us to the annoying problem of reactive power permanently oscillating, with double frequency, between source (generator) and sink (load). The author rejects this interpretation as being akin to the acceptance of a 'perpetuum mobile' in electrical circuits. The false perception that reactive power sloshes between source and sink without power losses contradicts any power flow calculations. As any engineer familiar with load-flow or power-flow calculations knows, at the end of such calculations we obtain reactive losses.

- In this author's interpretation, reactive power is not a physical entity existing only in AC circuits with reactive elements (reactors, capacitors, FACTS devices). Rather, it is the 'electromagnetic' momentum necessary to transport (potential) energy to the load; reactive power or momentum is the 'locomotive' or kinetic energy moving the energy train to its destination. For this reason, the so-called 'reactive' power is not an entity related only to AC circuits. As 'momentum', it exists in both AC and DC circuits. The difference is that in DC circuits there is only a linear momentum, whereas in AC circuits, there is a combination of a linear and a rotative momentum. The energy in AC circuits follows a helicoidal trajectory.

- This author rejects the concept of 'instantaneous' power as it contradicts Heisenberg's uncertainty principle.

- The author rejects the concept of 'electric' power, which is an electromagnetic phenomenon. Electromagnetic power, as well as electromagnetic energy, is perceived as more electric than magnetic, or more magnetic than electric, depending on the observer's position. *Mutatis mutandi*, on the basis of the theory of relativity, the author extends this interpretation to voltage and current. They are manifestations or reflections that depend on the observer's position.

- The author's physical interpretation is based on quantum electrodynamics and relativity.
- The author's mathematical representation is based on a non-commutative (non-Abelian) algebra and four-dimensional spacetime geometry. At the subatomic scale, this representation implies the transition from simple symmetry $U(1)$ to super symmetry $U(1) \times SU(2)$.
- The transition from the electric power concept to the electromagnetic power concept represents a break with the classical mechanics interpretation and a movement towards a quantum mechanics and relativity interpretation.

Appendix

Calculations

$$P_{\text{electromagnetic}} = (voltampère)\,(voltampère)^{\dagger} = (\bar{v} + \hat{i})\,(\bar{v} + \hat{i})^{\dagger} = (\bar{v} + \hat{i})\,(\bar{v} - \hat{i})$$
$$= (\bar{v} + I\bar{i})\,(\bar{v} - I\bar{i})$$

$P_{\text{electromagnetic}}$ – Electromagnetic power represents the space-time density of electromagnetic energy and electromagnetic momentum inside the circuit.

$\bar{v}$ – Voltage vector represents the electric field vector $(\bar{E})$, weighted by the permittivity (ε), inside the conductor (grade-1 multivector)

$$\bar{v} = \sqrt{\varepsilon} E$$

$\hat{i}$ – Current bivector (grade-2 multivector) represents the magnetic field bivector $(\hat{B})$, weighted by the permeability (μ), inside the conductor

$$\hat{i} = \frac{1}{\sqrt{\mu}}\hat{B}$$

The voltage vector and the current bivector are expressed in the same physical units, i.e.:

$$[\bar{v}] = [\hat{i}] = \sqrt{\frac{\text{Joules}}{m^3}}$$

I – Pseudoscalar (wedge product of orthonormal basic vectors in the Minkowski four-dimensional space, which is a pseudo-Euclidean space)

$$I = \gamma_0\,\gamma_1\,\gamma_2\,\gamma_3 = \gamma_0 \wedge \gamma_1 \wedge \gamma_2 \wedge \gamma_3$$

γ_0 – Time axis
$\gamma_1\ \gamma_2\ \gamma_3$ – Space axes

$$\gamma_0^2 = +1$$

$$\gamma_1^2 = \gamma_2^2 = \gamma_3^2 = -1$$

$$I^2 = -I^1$$

There are two geometrical products of interest:

$$(\overline{v} + \overline{Ii})(\overline{v} - \overline{Ii})$$

and

$$(\overline{v} + \overline{Ii})(\overline{v} + \overline{Ii})$$

The first geometric product will give us the electromagnetic power and electromagnetic momentum in the circuit, i.e. the spacetime density of electromagnetic energy. The electromagnetic energy is a sum of potential energy and kinetic energy. The electromagnetic kinetic energy is interpreted as electromagnetic momentum.

In other words, the author interprets electromagnetic power in circuit as the sum of two densities: 1) the density of the electromagnetic potential energy, and 2) the density of the electromagnetic kinetic energy, or the density of the electromagnetic momentum.

In classical electromagnetic theory, the sum of potential and kinetic energy is interpreted as an *invariant*: the Hamiltonian invariant. The author interprets the sum of the two densities as the Hamiltonian circuit invariant.

The result of the geometric product:

$$(\overline{v} + \overline{Ii})(\overline{v} - \overline{Ii}) = |\overline{v}|^2 + |\hat{i}|^2 + 2\hat{i}\overline{v} = |\overline{v}|^2 + |\hat{i}|^2 + 2\overline{Ii}\cdot\overline{v} = |\overline{v}|^2 + |\hat{i}|^2 + 2\overline{Iv}\cdot\overline{i}$$

is interpreted as a linear combination of a scalar and of a trivector.

The scalar part:

$$|\overline{v}|^2 + |\hat{i}|^2$$

is interpreted as representing, in the terminology of classical power theory, the *active power*.

The scalar part is a sum of two quantities:
$\frac{1}{2}|\overline{v}|^2$ – a scalar representing the power density of the electric field and,

---

[1] The pseudoscalar I is not to be confused with the scalar $i = \sqrt{-1}$

$\frac{1}{2}|\hat{i}|^2$ – a scalar representing the power density of the magnetic field

From a relativistic point of view, how much in an electromagnetic field is electric and how much is magnetic is observer-dependent. The vector part, expressed as:

$$2I\bar{v}\cdot\bar{i}$$

is a *trivector* representing the electromagnetic momentum density. It is interpreted as representing Poynting *vector* in classical electromagnetism.

The trivector is a grade-3 multivector, a geometric body endowed with translational and rotating movement; its trajectory follows an helicoidal pattern around the conductor ('wrapping' the conductor). The author interprets this trivector as corresponding to the concept of *reactive power* in classical power theory.

The author considers that the terms 'active power' and 'reactive power' are misleading; active power is a tautology and reactive power is an oxymoron. Both active power and reactive power are (electromagnetic) *power* and both potential and kinetic energy are *energy*. In recent publications on power theory, terms such as 'non-active power', 'non-active current' are currently used. The author of the monograph considers that such terms (including 'reactive power') lack physical justification and sound awkward. Reactive power is as much 'active' as the active power, both being electromagnetic power; 'inactive or non-active power' sounds like an oxymoron.

The second geometric product will give us two Lagrangians invariants: l_1 and l_2

$$(\bar{v} + l\bar{i})(\bar{v} + l\bar{i}) = |\bar{v}|^2 - |\bar{i}|^2 + 2(\bar{v}\cdot\bar{i})$$

$$l_1 = |\bar{v}|^2 - |\bar{i}|^2$$

is a scalar, representing the difference between the square of the electric power density and the square of the magnetic power density.[2]

The author interprets this expression as the *Lagrangian* of the circuit. It is similar to the mathematical expression:

$$E^2 - H^2,$$

that in classical electromagnetic theory expresses the difference between the square of energy of the electric field and the square of energy of the magnetic field.

$$l_2 = 2(\bar{v}\cdot\bar{i})$$

―――――――――――
[2] Slepian (Slepian, 1942) defined l_1 as reactive power

The above expression is also an invariant. The author interprets this expression as corresponding to the module of the Poynting vector.

The ratio of these two invariants:

$$\frac{l_1}{l_2}$$

gives us information about the Maxwell's field *complexion* (term used in modern electromagnetic theory) and is related to the velocity of electromagnetic wave. The author interprets this ratio as magnitude of the velocity with which the electromagnetic energy travels along the circuit.

In power engineering, this ratio appears to be related to the time difference between voltage and current known as the cos φ compensation problem. Interpreted as a ratio of invariants, the 'struggle' for the compensation of 'reactive' power could represent another 'ill-posed' problem.

$$\frac{l_2}{l_1} = \frac{2(\overline{v} \cdot \overline{i})}{|\overline{v}|^2 - |\overline{i}|^2} \cong \frac{2\overline{E} \cdot \hat{B}}{\overline{E}^2 - \hat{B}^2} \cong \arctan(2\theta)$$

The geometric product of the *voltampère* multivector with its reverse (conjugate) is related to the Lorentz relativistic force.

$$f_{Lorentz} = \frac{1}{2} \varepsilon_0 \left(\overline{v} + \hat{i}\right)\left(\overline{v} - \hat{i}\right) = P + \frac{1}{c^2} \text{ Poynting vector}$$

The author interprets the Lorentz's force as the equivalent to 'apparent' power, 'active' power as equivalent to electromagnetic power density and, the 'reactive' power as equivalent to electromagnetic momentum density.

Epistemology of Power Theory

1. Introduction

Between mathematical overdetermination and physical underdetermination, power theory suffers from an excess of mathematical formalism at the macroscopic scale and an insufficiency of physical knowledge at the mesoscopic and subatomic scales. Whereas modern physics rejects the classical model of an electron and asserts its particle-wave duality, textbooks on the circuit theory still retain the concept of current as a flow of (billiard-like) electrons. However, in the same textbooks, voltage and current are interpreted as waves and represented mathematically as trigonometric-valued functions. The main deficiency of power theory is the excess of its mathematization coupled with a poor understanding of the quantum-mechanical and relativistic nature of its physics. The excess of mathematical formalism that characterizes many publications on power theory reminds me of the warning given by Prof. Adolf Slaby (1849-1913): 'Don't let the science of electricity become a mathematical playground.' Like Hossenfelder (2018), while reading the power engineering literature I often get lost in mathematics.

2. Mathematical Guises and Disguises of an Elusive Physical Concept as Electrical Power

Motto: 'The pernicious influence of mathematics upon power theory' paraphrasing G.C. Rota, *Indiscrete Thoughts* [1997] (2008:xxi)

The mathematical representations of the concept of electrical power seem to recapitulate the development of the mathematical concepts of number and space.

The evolution of the *algebraic concept of number,* from real to complex and hypercomplex numbers, from scalar to pseudoscalar to oriented numbers, is mirrored in the representations of electrical magnitudes (voltage, current, active-, reactive- and apparent-power) as real numbers, complex numbers, and hypercomplex numbers).

The evolution of the *geometric concept of length,* from directed line to vector, oriented surface or bivector, oriented volume or trivector, multivector, spinor, etc., is mirrored in the representations of electrical magnitudes (voltage and current as vectors, apparent power as a multivector, reactive power as a bivector or trivector).

In the following pages, the author reviews the evolution of the mathematical representations of the electrical magnitudes active, reactive, and apparent power, as well as voltage and current.

2.1 Electrical Magnitudes Expressed as Real-valued Functions of Time; Power Equations as Partial Differential Equations: Bedell and Crehore

In numerous books and articles published between 1893 and 1927, Bedell and Crehore promoted the mathematical formalism of differential equations and the use of real algebra (R) for representing the power phenomena:

$$dw = eidt$$

$$eidt = \frac{idt \int idt}{C} + Ri^2 dt + Li\frac{di}{dt}$$

Bedell and Crehore (Bedell, 1927; Crehore 1893a, b) also promoted a time-consuming and cumbersome graphical approach, to which Macfarlane, Steinmetz, and Kennelly opposed analytical approaches based on hyperbolic quaternion algebra, complex algebra, and hyperbolic trigonometric functions, respectively.

2.2 Electrical Magnitudes Expressed as Complex-valued and Vector-valued Functions: Steinmetz and Janet

Motto: '...the shortest and best way between two truths of the real domain often passes through the imaginary one.' Hadamard, *The Psychology of Invention in the Mathematical Field* (1945:123).

From the first introduction of complex numbers, negative attributes like *absurd, meaningless, impossible, imaginary, perplex, hallucinatory, fictitious,* and *sophistic,* were applied to them (Nahim, 1998; Merino, 2006; Nikouravan, 2019). However, the imaginary j ($j^2 = -1$) remains one of the most successful mathematical oxymorons. Complex numbers were almost prohibited by the Spanish Inquisition. No wonder that the introduction of complex algebra in power theory received a lot of opposition! However, it is important to underscore that Thomas Johann Seebeck (1770-1831), Hermann von Helmholtz (1821-1894), and Adolf Franke (1865-1940; Franke, 1891) introduced complex numbers to the science of electricity before Kennelly (1893) and Steinmetz (1893).

Chapter 2 presented a detailed analysis of Steinmetz's symbolic method, which is still accepted as the working paradigm in circuits operating under sinusoidal operating conditions. This paradigm fails for three reasons. First, it relies on an antinomy between a commutative algebra (complex algebra) and an anti-commutative vector calculus. Second, it identifies complex numbers with (rotating) vectors. Finally, it assumes that double-frequency oscillations in time (i.e., power) are equivalent to a two-fold rotation in Argand's two-dimensional space. In hindsight, Steinmetz's symbolic method, based on a commutative complex algebra, represented a mathematical step backward compared to a non-commutative hypercomplex algebra, such as quaternionic algebra.

2.3 Electrical Magnitudes Expressed as Hypercomplex-valued Functions: Macfarlane and Kennelly

The hyperbolic numbers are relevant to power theory. Macfarlane (1893) used them in his power expression, and Kennelly (1928) used them in his calculation of electrical circuits.

The extension of real numbers with the new square root $u = \sqrt{1} \notin R$ leads to the concept of the hyperbolic plane H analogous to the complex number plane C. Both C and H planes represent two-dimensional vector spaces over the field of real numbers R. The hyperbolic numbers $H = R(u)$ represent an extension of the field of real numbers with the unipotent number u (just as complex numbers extend the field of real numbers with the imaginary i).

The Lorentz transformation replaces the Galilean transformation law (Weisstein, 1999).

A full treatment of the use of hyperbolic numbers in power theory is outside the scope of this monograph; however, the author investigated applications of hyperbolic numbers in power theory, signal processing, and electrical machines, including the contributions of the following: Y. Beck, N. Calamaro, and D. Shmilovicz (2016); M. Depenbrock and V. Staudt (1998); J.H.R. Enslin, and D.F. van der Merwe (1990); O.J. Ferguson (1903); A. Ferrero (1996); S.L. Hahn and K.M. Snopek (2017); R.S. Herrera and P. Salmeron (2007); A.E. Kennelly (1893); I.V. Lindell (1983); A. Macfarlane (1897); K. Muralidhar (2015); P.T.- Sah (1936); L. Serrano-Iribarnegaray (1993); J. Stepina (1989); C-T. Tai (1997); and J.L. Willems (1996).

2.4 Heaviside Operational Calculus and the Steinmetz Symbolic Method: Two Types of Mathematical Transformations in Circuit Theory

Ernest Julius Berg was a close co-worker of Steinmetz during their activity at General Electric in Schenectady, New York and later inherited Steinmetz's position as professor at Union College in the same city. Berg recognized the similarity between Steinmetz's symbolic method and Heaviside's operational calculus; both methods are considered as *symbolic* methods (Moore, 1966, 1968, 1970, 1971, 1995).

Operational Calculus owes its inception to Gottfried Wilhelm Leibniz (1695), who was surprised by the resemblance between two mathematical expressions: the expression (or the formula) for *n*-fold differentiation of a product of two variables and the expression (or the formula) for the *nth* power of a sum of two variables. This similarity permits the *algebraization* of a system of differential equations.

In his *Théorie analytique des probabilités* (1817), Pierre-Simon Laplace introduced the so-called *exponential shift* e^{-xp}; multiplication with this exponential shift *transforms* a real-valued function into a complex-valued function (Widder, 1945). As an example, the Laplace-Mellin *transform*:

$$F(p) = \int_{-\infty}^{\infty} e^{-xp} \, f(x)dx$$

assigns a function $F(p)$, of a complex-valued variable operator p, to a function $f(x)$ of a real-valued variable x. Essential, in a Laplace transformation (bijection), is that it *maps* (or 'sends') functions of real-valued variable onto functions of complex-valued variables. It permits solving problems posed in one mathematical domain into a problem-posing in another mathematical domain.

Lagrange carried this analogy further and gave the symbolic transformation formula:

$$f(x + h) = (e^{h\frac{d}{dx}})(f(x))$$

Boole, who contributed further to the theory of differential equations with constant coefficients, used the symbol D for the operation of differentiation.

Without knowledge of the previous contributions, Heaviside – an autodidact – developed independently a convenient method for solving differential equations. He was interested in solving differential equations, describing the transient phenomena in telephone and telegraph cables (electrical circuits for transmission of signals).

The main idea behind the above-mentioned mathematical transformations was to replace a system of differential equations with a system of algebraic equations and thus reduce the operation of differentiation to a simple multiplication.

For this reason, Heaviside introduced the operator p, which would take the expression x^n and transform it into nx^{n-1}.

Heaviside's *transform* is a *mapping* of a mathematical function (of a real-valued variable, e.g., a function of time) into an operator-valued function:

$$f(x) \rightarrow F(p),$$

where the operator p replaces the operation of derivation: $p \rightarrow d/dx$.

In his investigation of power theory, Steinmetz encountered a similar problem: how to solve the system of differential equations that model the behavior of electrical circuits for power transmission. These equations were usually solved through tedious and cumbersome numerical methods (Bedell and Crehore, 1892).

Steinmetz approached this problem as a *transform* problem. His transform consists in replacing a function of real-valued variable(s) with a function of complex-valued variable(s). His approach is similar to Heaviside's approach, in that both Heaviside and Steinmetz *transform* or *map* a mathematical problem from a real domain on to a *transformed* domain. The solution is found in the 'transformed' domain and then, through a back transformation, the results are 'translated' in the original domain.

The revolutionary idea of extending the concept of *number* in circuit analysis belongs to both Heaviside and Steinmetz. Steinmetz introduced the complex number, and Heaviside introduced the operator p.

Another common trait of these towering scientists is a highly intuitive and heuristic approach, sometimes devoid of mathematical rigor. This approach is evident in Heaviside's 'dinner' aphorism: "I enjoy the dinner, even if I do not fully understand the process of [mathematical] digestion."

Another similarity was that both were fiercely attacked by their more rigorous fellow mathematicians. Steinmetz was criticized for his assertion that $j^2 = +1$, which contradicts the tenet of complex algebra that $j^2 = -1$ (Besso, 1900). Heaviside was criticized for introducing a peculiar function (the step-function) that does not possess a derivative.

The methods of both Heaviside and Steinmetz use the exponential shift:

$$\text{Heaviside - } e^{pt}$$
$$\text{Steinmetz - } e^{-jwt}$$

Both methods predate Dirac's expression in quantum mechanics:

$$\phi' \rightarrow e^{\infty}\,\phi$$

Both methods correspond to an Abelian symmetry: $U(1)$

The exponential shift is present in Heaviside's method in the form $\phi(t) \to f(p)e^{pt}$ and in Steinmetz's method in the form $u \to Ue^{jwt}$ and $i \to Ie^{jwt}$ (Casper, 1926).

The exponential shift introduced by Heaviside and by Steinmetz would later be used not only by Dirac in the field of quantum mechanics, but also by Gabor (1946) in communication theory. It represents a global phase transformation:

$$\phi' \to e^{\infty} \phi$$

equivalent to

$$\psi' \to e^{jw} \psi$$

The exponential shift is also present, as a basic transformation, in Maxwell's electromagnetic theory:

$$\phi \to e^{i\lambda} \phi$$

The exponential shift reflects the $U(1)$ Abelian symmetry that characterizes classical (Maxwellian) electromagnetic theory. The well-known theorem of Emmy Noether (1918) established a fundamental connection between symmetries and conservation laws – a mathematical symmetry corresponds to a physical invariance. In classical electromagnetism, the conserved physical magnitude is the electric charge.

The task of *'rigorizing'* Heaviside was taken up by his followers, the Neo-Heavisideans – Berg (1924); Bush (1929); Carson (1927); Pol (1929, 1932) , and more recently, by Doetsch (1943), Mikusinski (1949, 1950, 1959); Jeffreys (1964); and L. Berg (1965).

Heaviside's operational calculus suffered from a non-rigorous mathematical foundation. Norbert Wiener (1926:559) states that "... the problem of obtaining a rigorous interpretation of Heaviside's theory... is still open." Yosida (1984:v) complains about the difficulties of understanding Heaviside's writings and considers that "the validity of his calculus remains unclear until now."

By the 1930s, G. Doetsch and many other mathematicians (e.g., Bromwich, 1928; Dalzell, 1930; Wagner, 1916, 1941) had begun to strive for a rigorous foundation of Operational Calculus, using a combination of algebraic and analytic methods. They used two different mathematical spaces – a space of the original function f and a space of its image, F. The two were connected through a *transform* (Laplace *transform*):

$$F(p) = \mathcal{L}[f]\,p = \int_0^{\infty} e^{-pt} f(t)dt$$
$$p \in C$$

The 'mysterious' Heaviside multiplication with the operator p is replaced by the multiplication with e^{-pt} in which the variable p is a complex number. The disadvantage of this approach consists in the fact that is a mixture of analysis and algebra.

A return to the algebraic methods was achieved by Mikusinski, who introduced the classical integral of convolution:

$$x(t) * y(t) = \int_0^t x(t-\tau)\, y(\tau)d\tau$$

This expression is also known as *'Duhamel convolution'*.

In his book *Operational Calculus*, Mikusinski gives a rigorous treatment of the calculus of operators together with many applications of the theory. The essence of the method consists of the concept of *convolution* of two functions, which are not necessarily continuous and real-valued. The convolution, defined by the symbol *, is not a multiplication. The convolution is both commutative and associative. In addition, in accordance with the Titchmarsh theorem,

$$f * g = 0$$

only if at least one of the functions is zero. This means that the quotient f/g always exists and defines an operation. Mikusinski also shows that an algebra of operators can be developed in an analogous way with the algebra of numerical fractions. Mikusinski's operators can be regarded as generalizations of both complex numbers and functions. Thus they reduce the task of solving ordinary differential equations with constant coefficients to routine algebraic operations.

Mikusinski provides a strict mathematical basis for Heaviside operational calculus without recourse to the Laplace transform theory.

Mikusinski's integral convolution has important applications in automatic control and signal theory.

The integral of *convolution* (or 'folding', or *Faltung* in German) is expressing the amount of overlap of one function $f(t)$ with another function $g(t)$ as the second function is shifted in time $g(t-\tau)$:

$$f(t) * g(t) = \int_0^t f(\tau)\, g(t-\tau)d\tau$$

It 'blends' one function with another function; however, it is not a multiplication. The mathematical operation of convolution has the following properties:

$$f * g = g * f \text{ (commutativity)}$$

$$f * (g * h) = (f * g) * h$$

$$f * (g + h) = (f * g) + (f * h) \text{ (associativity)}$$

The derivative of a convolution can be expressed as:

$$\frac{d}{dx}(f*g) = \frac{df}{dx}*g = f*\frac{dg}{dx}$$

The area under a convolution is the product of areas under the factors.

Mikusinski's integral of convolution is of interest because a number of authors studying power theory have approached the concept of electrical power from the viewpoint of signal theory, viewing electrical power as *the power of a signal*. Among them are Ghassemi (2004), Hartman and Hashad (2008), Jeltsema and Kaiser (2016), Krajewski (1990), Nowomiejski (1964, 1981), and Nossek and Ivrlač (2011). These authors used the *integral of convolution*, whereas the existing power paradigm uses the simple *multiplication of two functions* of time – voltage and current as functions of time.

The present author stresses the similarity between signal theory and circuit theory in dealing with the concept of power (e.g., power of a signal and electrical power in its real and complex expression):

$$f(t)*g(t) = \int_0^t f(\tau)g(t-\tau)d\tau$$

versus

$$p(t) = v(t)i(t)$$

$$\dot{S} = \dot{V}I*$$

In circuit theory, $v(t)$ and $i(t)$ are complex-valued functions of a real variable t; in signal theory $f(t)$ and $g(t)$ are complexified signals. As mentioned in the monograph, there is no difference between Steinmetz's transform and Gabor's transform; both are based on Hilbert transform. So why do we treat the power of a signal differently from electrical power? The power of a signal is obtained as a convolution, whereas the expression of electric power is obtained as a simple multiplication of voltage and current (although both could be interpreted as complexified signals)!

In modern signal theory, the power of a signal is interpreted as a monogenic signal (e.g., Felsberg and Sommer, 2001). A monogenic signal is a generalization of the analytic signal (a *one*-dimensional mathematical entity) to a *two*-dimensional entity. The analytical signal is based on Gabor's transform, whereas the monogenic signal is based on Riesz's transform, which is used in image processing.

However, other researchers (e.g., Lev-Ari and Stanković, 2004) investigating power theory from a signal-processing perspective used the following power expression:

$$<v,i> \triangleq \int v(t)i(t)^T \, dt$$

I wish to underscore the lack of agreement between specialists in signal processing and specialists in circuit theory with respect to the concept of power.

Further developments in Operational Calculus include generalization of the Fourier integral, Plancherel's L^2 theorem, Wiener's contribution to generalized harmonic analysis, and finally L. Schwartz's theory of distributions (Barany, A-S, Paumier, and J. Lützen, 2017; Lützen, 1982). The theory of distributions, which uses functionals, is mathematically a better reflection of the process of physical measurements. In practice, measurements coming from an industrial process are 'smeared' or corrupted by random errors, which intrude upon the estimation of the observables (e.g., voltages, currents) and make the process of estimation difficult. Instead of working with functions of variables, a functional works with the probability densities of the variables. As a consequence, distribution theory permits a more robust state estimation of industrial processes. A possible application would be the online and real-time state estimation of power system security.

2.5 The A-C Kalkül: Mathis and Marten

Basing their work on rigorous mathematical foundations (the modern Laplace transform, Mikusinski's integral of convolution, and the Yosida concept of hyperfunction, i.e., generalization of the function concept), Mathis and Marten achieved a theoretical breakthrough in power theory (Mathis, 1987, 1994; Mathis and Marten, 1989; Marten and Mathis, 1992; Küpfmüller *et al.*, 2013).

The A-C Kalkül (AC Calculus) not only represents an innovative application in power flow calculations, but also raises a fundamental question regarding power definition.

I distinguish between the physical concept of electrical power and the mathematical concept of electrical power. As correctly pointed by Czarnecki, power theory is a mathematical theory, and in the author's opinion, the concept of power is an epistemological, mathematical concept.

In this context, the A-C Kalkül confronts us with the choice between two 'axiomatic' definitions of power that are completely different:

- The classical definition due to Blakesley-Ferraris: power as a product of voltage and current (multiplication of two functions)
- The new definition due to Mathis and Marten: power as a convolution integral of voltage and current

Unfortunately, the important contribution brought by the A-C Kalkül is little known outside the domain of German technical literature (but *see* Gerbracht, 2007). It deserves a broader discussion. However, this topic is outside the scope of this monograph.

2.6 A Conjecture: Electromagnetic Power as Density of Electromagnetic Force in Circuits

I will examine the concept of power from the position of two modern physical theories: the theory of Restraint Relativity and the theory of Quantum Electrodynamics. This examination will redefine the concept of power. In this interpretation, power represents a density: the density of the electromagnetic force. The following paragraphs discuss the main aspects of the new theory.

2.6.1 Epistemological Aspects

This monograph uses the mathematical formalism of geometric algebra; however, it adopts a pluralistic mathematical approach. Other mathematical systems, such as differential forms (Cartan) or density functions (L. Schwartz) are considered as alternatives for expressing the concept of power.

In chapter 4, electromagnetic power is represented as a multivector in the relativistic four-dimensional spacetime manifold. Electromagnetic power (akin to 'apparent' power) is a linear combination of a scalar (akin to 'active' power) and a trivector (akin to 'reactive' power).

I reject the old paradigm, in which electrical magnitudes are expressed as trigonometric functions, complex functions, and vectors. Under the assumption that voltage and current should be expressed as trigonometric functions of a network's frequency (50 Hz or 60 Hz), the expression for apparent electrical power ($p = vi$) is also a trigonometric function, albeit containing components of double frequency (100 Hz or 120 Hz). I consider power components of double frequency to be merely mathematical artifacts. However, many textbooks consider these mathematical entities to be physical realities; they are interpreted as components of power flowing back and forth between the generator and the load. Moreover, the current paradigm considers that these power oscillations do not cause any network losses. This is tantamount to asserting the existence of a *perpetuum mobile* in power systems.

The mathematical expression of electrical magnitudes as complex-valued functions induces the misleading physical interpretation (reification) of 'apparent' power as a complex function, and similarly, the misleading physical interpretation (reification) of 'reactive' power as something 'imaginary'.

The mathematical expression of electrical magnitudes as vector-valued functions induces an incorrect mathematical interpretation of apparent power as the sum of a polar vector ('active' power) and an axial vector ('reactive' power).

This monograph interprets voltage and current as 'place-holders' for the electric and magnetic fields in circuits. The electrical field has a *polar* attitude, whereas the magnetic field has an *axial* attitude. They cannot be separated and they are not independent entities. They are, in fact, the two faces of the same electromagnetic field. For this reason, I conceive a new physical and mathematical entity – the **voltampère** multivector – that is a linear combination between the 1-grade volt multivector and the 2-grade ampere (bivector).

The new paradigm inherits from Steinmetz the idea of an algebraic-geometric unification. However, contrary to Steinmetz's confusion between complex numbers and vectors, it uses geometric algebra, which integrates both complex algebra and vector calculus.

This new paradigm rejects the symbolic method. Steinmetz's expression for complex power is based on the false assumption that, for double-frequency phenomena, $j^2 = +1$; in other words, Steinmetz assumed that the imaginary scalar is a function of the frequency (rotational speed).

Janet's mnemonic expression $\dot{S} = \dot{V}I^*$ is also rejected. It could be justified (although Janet did not know it) if we were to consider voltage and current as complex vectors in Hilbert vector space. However, I reject both the assumptions that voltage and current are separate physical entities and their mathematical representation as independent complex vectors.

2.6.2 Ontological Aspects

The new ontology inherits from Steinmetz the idea that voltage and current are merely epiphenomena. In fact, at the macroscopic level, voltage and current are merely blackboxes. The current textbook paradigm positions voltage and current at the center of power theory, whereas I consider power to be the primary concept; voltage and current are only its macroscopic observables.

Contrary to the existing axiomatic definition of power as the product of voltage and current, I assert the primacy of electromagnetic power. Power is not a result of multiplying voltage and current as stated axiomatically in the expression:

$$p = vi$$

On the contrary, voltage and current are the manifestation or expression of electromagnetic power. They appear as local observables and as the Janus' faces of electromagnetic power. I interpret voltage and current as an inseparable, intertwined physical and mathematical entity that expresses (as a unit) the electric-magnetic duality of electromagnetic phenomena.

The expression $p = vi$ considers that instantaneous power is calculated from the simultaneous measurements of voltage and current. This idea is flawed. It contradicts Heisenberg's uncertainty principle in quantum mechanics, which asserts that you cannot measure simultaneously and with the same precision the position and the momentum of an electron. *Mutatis mutandis,* this principle applies also to the measurement of instantaneous power based on simultaneous measurement of voltage and current.

In conclusion, power is not the product of voltage and current. In my new definition, electromagnetic power is the product of the **voltampère multivector** and its reverse (*cf. Appendix to chapter 4*).

Contrary to the existing paradigm, in which electrical power is interpreted as a time derivative of energy, I assert that electromagnetic power is the spacetime density of electromagnetic energy-momentum; i.e., energy-momentum at this moment and in this place (here and now).

The new paradigm is based on the following assertions:

- Electromagnetic power is akin to Faraday's force in electrical circuits and to apparent power. Apparent power is not just a mathematical entity useful in design of electrical apparatuses; it is endowed with a physical meaning.
- The scalar part of electromagnetic power is akin to the density of potential energy and to 'active power' in the old terminology.
- The trivector part of electromagnetic power is akin to the density of the electromagnetic momentum, to kinetic energy, to 'reactive power' (in the old terminology), and to the *Poynting vector*!
- There is no 'power' (active or reactive) oscillating back and forth with double frequency. There is no 'complex' or 'vector' power, and there is no more 'imaginary' or 'reactive' power. There is only electromagnetic power as the spacetime density of electromagnetic energy and momentum in matter (circuits).
- The transmission of power can be visualized as energy flowing on a helicoidal trajectory. The 'mechanism' of the energy transfer at the macro level is related to the helicoidal trajectory of the electromagnetic trivector as described in chapter 4.
- The transmission of power can be related to the interactions between the electromagnetic energy (high amplitudes, low frequency) produced by power system generators and the electromagnetic energy (low amplitudes, high frequency) of the matter. This assertion adopts Hestenes' hypothesis, which relates the forces involving the *Zitterbewegung* 'trembling motion' of the electron-positron pair and the emission of photons to *electromagnetic* forces. The mechanism of energy transmission at the mesoscopic level is modeled by Bostick (1991); Kanarev (2000); Consa (2018); Hestenes (2019); Johnson (2019) and Van Belle (2019), as follows:
  - A paired electron and positron, tightly connected on a plasma-like toroidal ring, move with the speed of light in opposite directions and with opposite chirality (left-hand or right-hand orientation).
  - The electron collides with the positron and 'gives birth' to two

photons. This collision is interpreted as the transmutation of electromagnetic matter (carried by the electron-positron pair) into electromagnetic force (carried by the massless photons). The two photons move with the speed of light on a helicoidal trajectory but in opposite directions.

o The two photons give birth to another electron-positron pair and the electromagnetic force is transmuted into electromagnetic matter.

o The process repeats.

I do not subscribe to this model because it implies (a) an electromagnetic plasma at very high temperatures achievable only in laboratory conditions (e.g., the CERN Large Hadron Collider) and (b) very high energy levels (e.g., TeV). As a power engineer, I find it difficult to imagine that turning on a 100-watt bulb would imply a process at mesoscopic level requiring energies of TeV magnitude and temperatures of the order of 6000 K.

In my interpretation, the massless photon *carries* the spin and the charge of the electron-positron pair through the matter. The collision of particles electron-positron occurs as a result of the energy produced by the generators (power system). As the effect of collision, a photon wave is produced. The massless photon, moving on a helicoidal trajectory, carries the power system energy through the transmission system at a subluminal speed that depends on the material characteristics of the electrical network. The *photon* is the *carrier* of the *electromagnetic force*.

Nature has four fundamental forces: gravitational, electromagnetic, strong, and weak forces. The weak and electromagnetic forces have been unified in an electroweak theory by Sheldon Glashow, Abdus Salam, and Steven Weinberg. Scientific efforts are still being made to create a single theory unifying electromagnetic, weak, strong, and gravitational forces (Grand Unified Theory – GUT). For this reason, I am cautious about the above interpretation of the mechanism of energy transmission. The power theory has still a lot of loose ends, such as magnetic current, gauge theory, topological aspects, and electromagnetic vector potential.

Further progress in power theory requires us to leave behind the traditional approach confined to macroscopic electrical magnitudes (voltage, current, power). Further progress is possible *if and only if (iff)* this investigation of the concept of electromagnetic power and of the process of power transmission is extended to the level of subatomic physics.

3. Power Theory at the Mesoscopic and Subatomic Levels

Investigation of power theory at the mesoscopic level requires knowledge of particle physics, physics of condensed matter (Fradkin and Palchik,

1996), plasma physics, quantum metrology (Piquemal *et al.*, 2017), topological electrodynamics (Rañada and Trueba, 2001), and gauge theories (Zeidler, 2011).

The first basic question I had to re-examine as I began this investigation was: What is electric current?

Like the majority of power engineers, I associate electric current with the motion of electrons. And indeed, at more than one hundred years' distance, the answer to this question is similar – electric current is a flow of electrons, albeit a quantized fluid.

In his article 'Electricity' for the 11th edition of the *Encyclopaedia Britannica* (1910-11): 9:192), Sir John Ambrose Fleming wrote: "The operation called an electric current consists in a diffusion or movement of these electrons through matter, and this is controlled by laws of diffusion which are similar to those of the diffusion of liquids or gases." This definition is important because it marks the demise of classical Maxwellian electromagnetic theory based exclusively on fields. Classical electromagnetic theory negates the existence of charged particles; a particle is considered merely as a discontinuity of the fields. Fleming's definition is also important because he adopts Drude's (1900) model of electrons as a gas penetrating the lattice of the conductor.

Quantum theory, under the term 'current', distinguishes negative, positive, neutral, and magnetic currents. Renton (1990) interprets electric current as a process of creation and annihilation of particles.

In modern quantum metrology, the ampere is one of the seven base units of the international system of units (SI) and is defined as follows: "One ampere is the electric current corresponding to the flow of (1.602 176 634 × 10^{-19}) elementary charges per second. The electric charge, or Coulomb, is defined as $C = As$ (ampere multiplied by second) (*Bureau International des Poids et Mesures*, 2019).

In this author's opinion, both definitions are merely phenomenological descriptions and not conceptual interpretations; both beg the question: What is the electron?

For a theoretical physicist, an electron exists as a physical entity when it can be calculated. For an experimental physicist, an electron exists as a physical entity when it can be detected. For an engineer, an electron exists as an engineering entity when it can be measured.

My problem-posing was different. I asked two questions: (1) What is the causal relationship between electrons and electromagnetic power? and (2) How is the motion of electrons related to the transmission of power or electromagnetic force?

I adopted the interpretation of electric current as a quantized fluid. However, neither Fleming's definition of an electric current nor the quantum metrology provides a causal explanation for the questions: What is electromagnetic power? How is the power transmitted from generator to the load?

3.1 Electrons and Positrons

The fact that an electron is a carrier of electrical charge does not mean that the electron is also a carrier of force. And indeed, it is not. The carrier of force is the photon (Mark, 2015).

Bunge (1950) characterized the electron as inexhaustible. In fact, he 'plagiarized' Lenin, who used the same epithet for the electron in his book, *Materialism and Empiriocriticism* (1909).[1]

What is the electron?

It is one of twenty-seven elementary particles discovered in particle physics (Renton, 1990; Peskin, 2019).

The electron is a form of matter – electromagnetic matter – and it has a twin brother, the positron. The twins have both the same mass and electrical charges of the same magnitude, the electron with a negative charge and the positron with a positive charge (Anderson, 1934). Each has, in addition to six degrees of freedom, another type of momentum, the spin, but the electron has a different spin from the positron. They rotate with opposite chirality. The electron is a stable particle, while the positron has a very short lifetime which is measured in picoseconds.

The electron is not alone; it has a family composed of electron neutrino - v_e, muon neutrino - v_μ, tau neutrino - v_τ, muon - μ and tau - τ. The electron family is part of a larger family, the leptons, and the leptons are part of another particle family, the fermions. Dozens of books discuss the discovery of the electron and its 'biography' (e.g. Buchwald and Warwick, 2001; Arabatzis, 2006). Below is a brief summary of the most relevant information regarding this particle.

Experimental observations show that the electron's rest mass m_0 is 9.109382×10^{-31} kg, its electric charge c is $1.60217648 \times 10^{-19}$ Coulomb, and its spin (self-rotation), discovered by Goudsmit and Uhlenbeck is ½ $\hbar$. The size of an electron is less than 10^{-15} cm, and yet it is a huge size compared with that of Planck's quanta, which is 10^{-33} cm.

Since the 1890s, many different models of the electron have been proposed. These models include Thomson's plum pudding model (1897), Drude's cloud model (1900), Abraham's point model (1903), Lorentz's deformable electron introduced in 1904 (cf. Lorentz, 1916), and Rutherford's (1911) planetary model of the nuclear atom, consisting of a condensed atomic nucleus positive charge encircled by planetary electrons.

---

[1] In *Materialism and Empiriocriticism* (1909), Lenin expressed the view that the electron is as inexhaustible as the atom, and that nature is not only infinite, but exists infinitely. At the time of the book's publication, only the electron and the proton had been discovered.

Niels Bohr's (1913) model combines Rutherford's model of the nuclear atom with Planck's idea of a quantized nature of the radiation process. In this model, electrons orbiting on a circular orbit inside the atom are in stationary states. They do not radiate energy unless they 'leap' from one orbit to another. Arnold Sommerfeld's 1915 model extended Bohr's idea, but the electrons moved on elliptical orbits, and Parson (1915) introduced his ring model of the electron equivalent to a tiny magnet. Pauli in 1925 postulated the exclusion principle: that no quantum state can be occupied by more than one electron. Dirac (1928) suggested a point-like electron, and Felix Bloch (1929) developed a detailed wave-mechanical theory of electrons bound in atomic lattices (*see* also Bloch and Nordsieck, 1937). More recently, we have the 'amoeba' model of Bohm (1951), Schönfeld's (1990) spherical loop model, Bostick's (1986) plasma toroidal model, the topological model of Rañada (1992), the semi-classical models similar to the Large Hadron Collider of Bostick (1991), Consa (2018), Johnson (2019), and Kanarev (2000), and Hestenes' (2019) *Zitterbewegung* model.

A NASA report by Cambier and Micheletti (2000) contradicts the semi-classical model of Consa and Kanarev, stating that the electron spiral toroid concept is mathematically unstable and experimentally unsubstantiated. Nevertheless, the Consa and Kanarev model was later adopted by Van Belle (2019) and Johnson (2019).

Still other researchers have proposed different models, assuming that the electron is a photon trapped in a vortex (Bergman and Wesley, 1990; Williamson and van der Mark (1997); Robinson, 2011; Akins, 2015; Garrigues-Baixauli, 2019). According to Wilczek (2013), there is no evidence that the electron has an internal structure.

Bunge (1950:116) stated, "The question of the structure of the electron was declared devoid of physical meaning."And Russian physicist Yakov Frenkel demanded that all attempts to create a model for the electron be abandoned. He considered that any classical model is futile because the electron, as a quantum mechanical entity, cannot be represented in classical models.

In my opinion, the debate about how to model the electron is still open. It is reminiscent of the remark by American mathematician Oswald Veblen: "… people used to say that a physicist thinks of an electron as a particle on Mondays, Wednesdays and Fridays, and as a wave on Tuesdays, Thursdays and Saturdays, and on Sundays prays for a Messiah, who will lead him back to the belief which he held on Monday" (Veblen, 1934: 415).

High school and university physics lessons led me to perceive the electron as moving with the speed of light. In reality, the electron has a very fast but jerky, oscillating movement (*Zitterbewegung*, a highly oscillatory microscopic motion at the speed of light, as predicted by Schrödinger) with a very high frequency (1.23559×10^{20} Hz). However, its translational motion proceeds at a snail's pace of one mm/hour. With such a slow speed, it is difficult to explain the almost instantaneous transmission of electrical power through the high-voltage grid.

The movement of the 'enigmatic' electron (Wilczek, 2013) is brought to an end through collision with the positron. They both transform into the 'problematic' photon (Vistnes, 2013).

At the end of this short presentation, many questions regarding what the electron is remain unanswered: 1) What is the origin of the electron's charge? 2) What are its dimensions? 3) What is the spin? 4) Is an electron always a particle, a wave, or both? 5) Are electrons oscillating photons?

Despite my intensive search in the literature, a theory for the structure of the electron remains elusive (Robinson, 2011), and the electron remains an abstract thing. My hypothesis is that multiple models of the electron can co-exist, depending on the electron's energy level or frequency. I doubt that any of the existing models are adequate for the process of electrical power transmission at 50 Hz (with low levels of energy, but large quantities of power).

3.1.1 The Positron

The positron is not a stable particle (Roth, 2007). It has one of the shortest particle lifetimes, on the order of 159-366 picoseconds (ps). One picosecond is equal to 10^{-12} seconds. The relationship between one picosecond and one second is the same as the relationship between one second and 31,689 years! Just as an electron rotates in one direction with respect to its magnetic field, so the positron rotates in the opposite direction with respect to its magnetic field. Their spins are anti-parallel. While the electron is electromagnetic matter, the positron is considered anti-matter. An experimentally verified mathematical model that would precisely describe the attraction between an electron and a positron does not yet exist (Dorn, 2009, 2015). However, it is proved that their collision and mutual annihilation create the 'problematic' photons (gamma rays). This is a process of transmutation from matter into energy radiation:

$$e^+ + e^- \rightarrow \gamma + \gamma$$

The first suggestion of a positive particle came from P.A.M. Dirac in 1931. In 1932, C.D. Anderson from the California Institute of Technology discovered the positron, and the following year P.M.S. Blackett and G. Occhialini from the Cavendish Laboratory in Cambridge, UK, confirmed its existence(Chadwick *et al.*, 1933).

3.1.2 The Photon

The question "What is the photon?" is equivalent to the question "What is light?"

This is a difficult question because 'light is heavy' and the photons are supposed to be massless. According to the physics of particles, only leptons have mass, and the photon is not a lepton, but merely a massless boson carrying electromagnetic effects over human-scale distances.

The question: Do photons have mass? was the subject of a debate between Einstein and Niels Bohr. Einstein's point of view was that photons do have mass. He based his opinion on two arguments: 1) that light exerts pressure on objects, and 2) that light is bent under the effect of gravity.

Einstein was right, and this was confirmed by experiments of R.V. Pound and J. L. Snider (Pound and Snider, 1965). The fact that light exerts a force on objects is also demonstrated by the fact that a laser ray exerts a push force.

Other points of view are that photons could behave either as waves or as particles, depending on their energy level (Renton, 1990; Kracklauer, 2015; Rangacharyulu, 2015).

Renton states that at a value of $10GeV^2$ level of energy, the e^+ and e^- annihilation results in photons as electromagnetic waves, whereas at the $2000\ GeV^2$ level of energy, photons interact with matter, behaving as particles (Renton, 1990: 5).

I adopt the phenomenological point of view of M.B. van der Mark (2015c), which is that the photon is the transfer of a single quantized amount of energy and angular momentum between an electromagnetic transmitter and an absorber.

This transfer represents the process of power transmission at the mesoscopic scale, mediated by the photon as carrier of energy and momentum. I interpret the photon as a *process*, i.e., neither a particle nor a wave. In this process, the photon is a carrier of energy and momentum from the electromagnetic source to the electromagnetic sink.

This process of transmutation from electron into photon is represented musically by Richard Strauss' tone poem, *Tod und Verklärung* (Death and Transfiguration) – an appropriate metaphor for the unending process happening at both cosmic and microscopic scales. I recommend this highly emotional musical work – it is a good mental therapy for the reader trying to recover from digesting this monograph.

The wave model of the photon was introduced by Louis de Broglie (1924) to describe the phenomenon of light. The year 2015 was declared by UNESCO as 'The Year of Light'. However, light has yet to dawn. From my survey, which included the recent works and models of Hestenes, Dorn, Robinson, Consa, Johnson, and van Belle, I conclude that even today there is still no consensus among physicists about how to model the photon. It has been a mysterious object for centuries and still remains difficult to understand as a physical entity. However, I am inclined to interpret the photon as a wave carrying electromagnetic force.

The next question is: What are the main entities in quantum physics?

Modern Physics recognizes four major forces in Nature: 1) gravity, 2) electromagnetic force, 3) weak force and 4) strong force. The Standard Model[2] (Tanabashi et al., 2018) relies on seventeen fundamental particles and about two hundred non-fundamental particles, as well as these four physical forces.

Mathematically, I interpret the term 'force' as the gradient of energy. According to van der Mark (2015a: 1), "At astronomical sizes the attraction force is provided by gravitation, whereas at molecular and atomic sizes it comes from electromagnetism. For nuclear scales it is provided by the weak and strong interaction."

The strong forces dominate the weaker ones. The action of the weakest force – gravitation – is exercised at the longest distance and at the cosmic scale; the action of the strongest force – strong force – is exercised at the shortest distance and at the smallest subatomic scale. Weak forces act at distances of 10^{-18} m. Strong forces act at even shorter distances.

Forces are carried by bosons. Gravity forces are carried by gravitons, electromagnetic forces are carried by photons, weak forces are carried by $W^{\pm}$ and Z_0 bosons, and strong forces are carried by gluons.

Strong and weak forces are characterized by high levels of energy (*TeV* ÷ *GeV*); electromagnetic and gravity forces are characterized by low energy levels. The distinction between these forces is based on the concept of symmetry. Electromagnetic forces belong to the class of $U(1)$ symmetry (Abelian or commutative symmetry), and the weak forces have non-Abelian $SU(2)$ symmetry. The combined electromagnetic and weak forces (electroweak) are characterized by $SU(2) \times U(1)$ gauge symmetry, and the strong forces, by $SU(3) \times SU(2) \times U(1)$ gauge symmetry.

Table 5.1: Comparative Strength of the Forces of Nature

	Relative Strength
Strong nuclear	1
Electromagnetic	10^{-2}
Weak nuclear	10^{-15}
Gravitation	10^{-41}

The mathematical symmetry is important because it reflects a physical property – the conservation of a specific physical entity.

---

[2] 'Standard Model' is a quantum-mechanical description of all known elementary particles: from quarks inside protons and neutrons to fermions, leptons, hadrons, and bosons.

4. Power Theory – A Gauge Theory

In the following paragraphs, I will discuss 1) the meaning of gauge theory, 2) power theory as a gauge theory, 3) mathematical symmetries and physical conservation laws, and 4) the impact of gauge theory on power system analysis.

The meaning of the gauge theory is illustrated by the flock of birds resting (and linemen working bare-handed) on live, high-voltage transmission power lines (IEEE, 2003). For someone unacquainted with the gauge principle, this seems like a dangerous act; however, the electrical workers and the birds know better. The voltage on the transmission line is not absolutely high; it is high only in relation to the Earth's potential.

The pigeon can sit as safely on a high-voltage wire as on the ground, regardless of the wire's voltage, as long as there is *no electric* potential difference between the pigeon and the wire.

A gauge is a measure, and to gauge is to calibrate. The history of this term and its use in physics is 'very roundabout' (Weinberg, 1976:22) or circuitous. Here I use the term 'gauge theory' to characterize a full class of physical theories that manifest a principle of invariance as expressed in an internal (not spatial) mathematical symmetry. Gauge theories offer the prospect of unifying the weak, electromagnetic, and strong forces in one unified theory.

Defining a theory as a gauge theory is also a mathematical statement – it states that the theory is mathematically redundant in its description of physical reality. A gauge theory is any theory with an excess of mathematical structure; for example, classical electromagnetic theory and classical power theory are gauge theories. A theory is a gauge theory if it exhibits gauge freedom. Gauge freedom is associated with mathematical underdetermination (i.e., more than one mathematical solution). In other words, the mathematical structure exceeds the physical structure; the problem has a 'surplus mathematical structure'.

This situation occurs very often in physics. The mathematical description of any physical state or physical phenomenon contains excess degrees of freedom, so the same physical situation is equally described by many equivalent mathematical structures (as a very talkative person describes the same event again and again in different words). Physical theories tend to be gauge theories, i.e., theories in which the physical system is described by more mathematical variables than there are physically independent degrees of freedom.

An example from electrical engineering is the use of complex numbers in AC circuit theory. The physical circuits and their magnitudes are mapped on to the complex plane, yet, only the real part of the complex plane is necessary. The use of complex algebra is merely a mental crutch to facilitate the calculations.

Gauge theory is best understood geometrically. The gauge freedom is a consequence of the freedom of choice of coordinates in some space. Gauge theory is a type of field theory in which the Lagrangian function is invariant under a continuous group of transformations.

The classical theory of electromagnetism is a gauge theory belonging to an Abelian gauge group. Its gauge symmetry is a global symmetry (which does not depend on spacetime), and it is related to the conservation of electric charge and current.

A local symmetry is one where the symmetry group is continuous and depends on spacetime; such a gauge symmetry introduces a gauge field to the theory, i.e., a field that mediates a force. In the case of quantum electrodynamics, this local gauge symmetry is represented by the field of bosons in which the photon mediates the transfer of electromagnetic force or electromagnetic momentum. Neither classical electromagnetic theory nor classical power theory is endowed with such a gauge symmetry. My hypothesis is that both the theories lack conservation of electromagnetic momentum, and further, that the algorithm of load flow is incorrect with respect to the calculation of reactive power akin to the electromagnetic momentum.

Maxwell's equations display a symmetry between the electric field $\bar{B} \to \bar{E}$ and the magnetic field $\bar{E} \to -\bar{B}$, which expresses the duality between the two fields. The electric and magnetic fields are invariant under the following electromagnetic transformation (Saa, 2011):

$$E \to E' = E \cos \theta - B \sin \theta$$

$$B \to B' = B \cos \theta + E \sin \theta$$

A similar duality exists in power theory between the voltage and the current equations. At a deeper (relativistic) level, the above equations tell us that the values of E and B are observer-dependent. The perception of an observer 'at rest' with respect to the field E will be different from that of an observer 'at rest' with respect to the field B.

The 'space' of power theory is a topological space with the structure of a circle. It corresponds to the complex numbers of modulus 1, representing the electromagnetic matter phase at a point.

The freedom to choose many different potentials that describe the same electromagnetic fields is called *gauge invariance*. This gauge invariance is expressed in the invariance of the Lagrangian, which leads to the conservation of electric charge (Noether's theorem).[3] The Lagrangian is defined as the difference between potential and kinetic energy.

---

[3] Emmy Noether (1882-1935), a German mathematician at the University of Göttingen, stated in her theorem that for every symmetry of nature there is a corresponding conservation law, and for every conservation law, there is a symmetry.

In classical electromagnetism, the field strengths E and B are regarded as basic physical entities, the scalar potential φ and the vector potential A are introduced as convenient calculus devices, and the potential (φ, A) is not uniquely defined. Because of this gauge ambiguity, Maxwell's equations contain too much information and permit multiple solutions. Classical electromagnetic theory requires an additional constraint to produce a unique solution. Quantum Electrodynamics introduces an additional field, the massless gauge field of photons, which reflects the interaction between matter and radiation (missing in both classical electromagnetic theory and classical power theory).

By changing the potential (φ, A), we change the values of the electric and magnetic fields (E, B). This is a global gauge symmetry that does not express an instantaneous action at a distance. The idea of instantaneity contradicts the letter and the spirit of the theory of relativity, according to which the time delay between cause and effect is limited by the speed of light. Classical electromagnetic theory, which is based on Abelian (commutative) global gauge symmetry, contradicts the relativity theory. Therefore, it must be replaced with Quantum Electrodynamics QED, in which, in addition to the global gauge symmetry, we also have a local gauge symmetry that is non-Abelian (i.e., non-commutative). This local gauge symmetry reflects the transmutation of electromagnetic matter to energy via the boson that carries force: the photon.

In a nutshell, the question this author asks himself is this: What is the link between mathematical (analytical) symmetry and physical conservation in the case of both classical electromagnetism and classical power theory?

Both the theories possess an Abelian (commutative) symmetry expressed in the well-known expression:

$$\phi = \phi' = \phi e^{j\omega t}$$

Or in another form, the global rotation could be expressed as:

$$\psi(x) \rightarrow e^{i\theta}\, \psi(x)$$

The multiplication with $e^{i\theta}$ is called Weyl's gauge transformation and involves, at every spacetime point, a so-called $U(1)$ rotation, essentially a simple rotation in the complex plane.

This is a global transformation that keeps the total electrical charge invariant. The transformation, described by a change in the angle of rotation, leaves the relative phases between wave functions unchanged. Power engineers, who are familiar with load-flow calculations, know that changing the phase angle of the reference-bus voltage will not change the angle differences between the nodal voltages, so that active and reactive power flows and total losses remain the same (i.e., we obtain the same solution of the load flow calculations). This symmetry is akin to the symmetry of a circle and is described mathematically as $U(1)$ Abelian or commutative symmetry.

The symmetry group $U(1)$ is perhaps the simplest. It corresponds to the group of all phase factors, i.e., complex numbers of unit magnitude.

This analytical symmetry corresponds to a physical conservation law: the total electric charge carried by the electrons is kept constant. Mathematically, the Lagrangian of the system is also kept invariant. This *gauge* symmetry is a global symmetry: that is, by changing the phase of the voltage at one node (the reference bus), we rotate or shift the phase of all voltages by the same angle. As a result, the solution of the load flow calculations will remain (electrically) the same. The $U(1)$ gauge symmetry ensures the conservation of the electrical charge.

Heaviside and Hertz modified Maxwell's electromagnetic theory by eliminating the electromagnetic vector potential. Power theory is similar in its neglect of the electromagnetic vector potential. In short, it ignores the electromagnetic momentum. My claim is that the current power theory and the load-flow algorithm corresponding to it do not satisfy mathematically the physical law of electromagnetic momentum conservation.

Therefore – and this is one of my main proposals – the future algorithm of load flow should include a more profound symmetry, the local non-Abelian symmetry $SU(2)$. This additional gauge symmetry is equivalent to the introduction of an interaction with an additional field – the bosonic field of photons. This additional local gauge symmetry will ensure conservation of the electromagnetic momentum, and it takes into consideration the vector electromagnetic potential (neglected in the classical power theory). Therefore, the global Abelian symmetry $U(1)$ ensures the invariance of the electric charge carried by the electrons (particles with mass), whereas the local non-Abelian symmetry $SU(2)$ ensures the invariance of the electromagnetic momentum or forces carried by the massless photons. $U(1)$ is akin to the symmetry of a circle (planar connectivity), while $SU(2)$ is akin to the symmetry of a sphere (spatial connectivity).

4.1 Power Theory and the Physics of Condensed Matter

The phenomenon of power transmission at the macroscopic scale can be definitively elucidated only when we understand the phenomena taking place at the mesoscopic[4] level, i.e., the transmutation of electromagnetic

---

[4] A mesoscopic system is characterized by sizes varying from nanometer to micrometer; at this size, new phenomena are discovered (e.g., mesoscopic resistances in series do not follow the simple addition rules observed at macroscopic size)

matter to energy momentum to electromagnetic matter and so on. Maxwell's classical electromagnetic theory, conceived as operating in a vacuum at the macroscopic level, is not applicable at mesoscopic scales of condensed matter. The study of power transfer at mesoscopic scales requires knowledge of quantum electrodynamics, physics of condensed matter, and particle physics. In the mesoscopic domain, in addition to electromagnetic force, the weak force is acting, and the mesoscopic domain is marked by a symmetry breaking.[5]

The transition from the macroscopic to the mesoscopic domain is characterized by the phenomenon of "spontaneous symmetry breaking in gauge theories" (Kibble, 2015).

4.2 Power Theory and Quantum Metrology

As an engineer, I consider that we can understand a physical entity or a physical phenomenon when it is observable and measurable. According to Einstein, this is not always true because it is the theory that defines what we observe and what we measure.

For this reason, I investigated the literature related to quantum metrology and to measurements of voltage, current, power, and the fundamental electrical constants. I was surprised by the fact that, in quantum metrology, the measurement of macroscopic electrical power is based on the Planck constant (Stock, 2011:3940). All electrical units are nowadays redefined and based on quantum phenomena (e.g., the Josephson and Hall effects). The definition of current is based on the electron's elementary charge. A watt balance establishes a relationship between a macroscopic mass and the behavior of the microscopic world quantized by Planck's constant $\hbar$ with an accuracy of 5 parts in 10^8 (Olsen *et al.*, 1989).

The electric current is defined as a 'flow of electrons' measured in amperes; one ampere is the flow of one coulomb of charge per second:

$$1A = 6.25 \times 10^{18} \text{ electrons}$$

The ampere is one of the seven base units of the International Systems of units (SI), and in a completely different approach from the past, is defined on the basis of physical quantum effects in condensed matter (physics of solids). Quantum effects ensure that physical quantities, unlike their classical behavior, take only discrete values. There are discussions about the realization of a *quantum ampere meter* based on the fact that since

---

[5] An example of symmetry breaking is the phase transition of a liquid from room temperature to the frozen state; at low temperatures, we experience the transformation of the homogeneous medium into a non-homogeneous medium having a different spatial orientation

the late 1980s, it has been possible to manipulate single electrons (Ahlers and Siegner, 2011; Giblin, 2011).

In conclusion, the fourth industrial revolution – the quantum revolution – has opened new frontiers in metrology of electrical units; a quantum power metrology is still in waiting.

5. Conclusion

Knowing *what* happens is not the same as knowing *how* or *why* it happens. The classical power theory still remains a mathematical descriptive theory – a phenomenological theory. In order to understand power phenomena at the macroscopic scale, which are characterized by magnitudes on the order of 10^9 watts, 10^6 meters and seconds, we have to understand power phenomena at the mesoscopic scale, with magnitudes on the order of 10^{-18} watts, 10^{-15} meters, and 10^{-12} seconds.

There is a dialectical relationship between the macro-world and the micro-world; indeed, in order to understand the macro-phenomena, we have to understand phenomena at their micro-scale and be cognizant of quantum mechanics and relativity theories. At present, in my opinion, the prevailing theory of electrical power is approximately 80 years behind the developments in physics.

Electromagnetic power theory will be transformed from a merely mathematical and phenomenological descriptive theory into a conceptional and causal interpretative theory only by being part of the fourth industrial revolution: the nanotechnological revolution. In the meantime, power engineers should realize that power theory is still a theory 'under construction'.

The current power theory is ambiguous with respect to the wave-particle duality. All third-year students treat voltage and current as waves; at the same time, they conceive power transmission as a phenomenon akin to the motion of electrons. Any power engineer will explain the almost instantaneous transfer of energy by referring to electrons moving at the speed of light. In reality, the translational speed of electrons is very slow (only mm per hour). They have a rapid oscillatory movement (*Zitterbewegung*) with high-frequency oscillation $\cong 10^{20}$ Hertz. What is 'moving' with the speed of light is not the electron, but the photon, the carrier of electromagnetic force. The textbooks in power engineering do not even mention the existence of photons.

The speed of power transmission is related to the fact that the (almost massless) photons transport energy with the speed of light. The electrons are carriers of electromagnetic matter while the photons are carriers of electromagnetic force. Power transmission is a continuous process of transport of quantized energy as follows:

> Collision of an electron and a positron (carriers of electromagnetic matter) → annihilation of electromagnetic matter → transmutation in electromagnetic energy-momentum → photon (carrier of electromagnetic force) transportation at the speed of light → transmutation in electromagnetic matter carried by electron-positron pair →...

In my model, the electrons are behaving as particles and the electric current is a quantized flow of electrons and positrons traveling in opposite directions, while the photons are behaving as waves. The electrons (leptons) are carriers of electromagnetic matter. This fact supports my definition of power as *electromagnetic power*. The photons (bosons) are carriers of electromagnetic force (the gradient of electromagnetic energy and momentum). The photons interact with other bosons ($W^{\pm}$, Z_0) that are carriers of a weak force.

The electromagnetic forces and the weak forces are unified in the theory of electroweak forces – a topic that is outside the scope of the monograph.

Epilogue as Prologue

Motto: In most sciences, one generation tears down what another has built. In mathematics alone, each generation builds a new storey to the old structure.
— Haenkel, 1869

1. Can We Unify the Concepts of Power in Circuits and Energy-momentum in Electromagnetic Fields?

The classical theory of power in AC circuits is incompatible with the classical theory of electromagnetic fields. The antinomy resides in different axiomatic definitions of power as active, reactive, and apparent power in electrical circuits, and as energy-momentum in electromagnetic fields.

The present investigation started from the assumption that there is a mathematical similarity between Steinmetz's expression of complex power in circuits and the expression for energy-momentum. And indeed, when they are formulated in the mathematical formalism of geometric algebra, there is a striking similarity between Steinmetz's power expression:

$$v \cdot i + v \wedge i,$$

where $v \cdot i$ represents active power and $v \wedge i$ represents reactive power, and

$$E \cdot H + E \wedge H,$$

where $E \cdot H$ represents electromagnetic energy and $E \wedge H$ represents the electromagnetic momentum (and the Poynting vector).

As discussed in Chapter 3, power theory is rooted in Amperian electrodynamics, which derives the concept of force from the interaction of two wires (carrying-current, i.e. charges), whereas electromagnetism considers charge as a discontinuity in the field and force resulting from the interaction of electric and magnetic fields. Power theory and classical electromagnetic theory cannot be reconciled as they are axiomatically different.

Both the classical circuit theory and classical electromagnetism have been developed within the framework of classical mechanics; no 'bridging' between circuit theory and field theory is possible within classical mechanics. However, the monograph offers another perspective: it shows that bridging of the two theories is possible within the framework of quantum electrodynamics and restraint relativity.

2. Is the Current Power Paradigm Still Valid?

Electric power is currently expressed in many mathematical formalisms, such as real algebra, complex algebra, vector calculus, trigonometric algebra, quaternionic algebra, geometric algebra, and algebra of hypercomplex numbers. However, none of these mathematical theories reveal the physical reality.

I recommend the demise of Steinmetz's symbolic method (and of Janet's heuristic expression $S = \dot{V}I^*$).

My peers will not take lightly my accusation that the symbolic method of Steinmetz and Janet's heuristic expression $S = \dot{V}I^*$, in use for more than 120 years are fraught with mathematical and physical inconsistencies. Although I consider myself an 'honest' heretic,[1] I cannot wait until the last epigones of the expression $p = vi$ are safely in their tombs. The time has arrived for a rigorous debate regarding the concept of power and the mechanism of its transmission.

The main arguments for the demise of the current power theory (put forth in Chapters 2 and 4) are the following:

- The concept of instantaneous power contradicts the relativity theory. Power, voltage, and current are not causally related; they are correlated. Voltage and current are just manifestations of power
- The idea that we can measure voltage and current simultaneously and with equal precision contradicts Heisenberg's uncertainty principle
- Voltage and current are neither independent nor separable physical entities
- Voltage and current are *conjugate* representations of the concept of power. For this reason, it is incorrect to represent them mathematically with the same symbol (be it complex number, vector, or quaternion)
- Steinmetz derives the expression $P + jQ$ on the basis of the faulty assumption that, for double-frequency oscillations, the square of the imaginary equals one: $j^2 = +1$. In Chapter 2, I rigorized Steinmetz's

---

[1] Every honest heretic advances the scientific truth, provided he is not burned first

symbolic method and demonstrated that, in fact, the results are mathematically correct if expressed in the mathematical formalism of geometric algebra

- Janet's ubiquitous expression for complex power $S = \dot{V}I^*$ (often considered as equivalent to Steinmetz's expression for power) is correct if and only if voltage and current are represented in Hilbert complex vector space. In a sense it is ironic that Steinmetz rediscovered Grassmann-Clifford geometric algebra and that Janet used the rule for multiplication (inner product) of complex vectors many years before Hilbert introduced his 'new' algebra of complex vector space
- The current textbooks use indiscriminately trigonometric, complex, and vector representations of electrical magnitudes

o *Trigonometric representation*

Representing voltage and current as harmonic plane waves is conducive to a number of misconceptions; for example, active power is interpreted as DC power. Similarly, a part of the active power is interpreted as a double oscillating component. The most egregious misconception is the representation of reactive power as a double-frequency entity, oscillating permanently between generation and load without producing any network losses. This is nothing less than a tacit acceptance by power engineers of the *perpetuum mobile* principle, which has long been dismissed.

It is a shame for our discipline that textbooks and articles canonize the idea of reactive power sloshing back and forth between source and sink (or being 'lent' to the load for half a cycle and 'returned' to the generator in the second half of a cycle). It is even worse – the IEEE Standard 270-2006 defines reactive power as a quantity that flows back and forth in an AC circuit without being consumed! It raises the question: Why are utilities charging consumers for a quantity that it is not even consumed?

Fryze (1932: 625)[2] long before me and more recently Czarnecki (2020) have repeatedly criticized the misconception that reactive power oscillates in electrical systems without causing any losses. In addition, any engineer familiar with power flow calculations knows that at the end of those calculations, we obtain results that include reactive power losses. Industrial consumers pay huge penalties for reactive power consumption. In the energy market, reactive power is traded like power; it is inconceivable for an economist to trade something that is not consumed and nevertheless to make profit of it (unless it is a Ponzi scheme).

---

[2] S. Fryze: "Die irrige Meinung, dass die Blindleistung an das Pendeln der Energie gebunden ist" (Author's translation: The erroneous opinion that reactive power is related to energy oscillations)

○ *Complex representation*

Representing electrical magnitudes as complex numbers creates a number of mathematical ambiguities:

- Voltage and currents are both complex numbers; however, we cannot (and we should not) add them
- Apparent power as a complex number is void of physical meaning; it is wrongly interpreted as a mere calculation device
- Reactive power receives a non-physical interpretation (something that is 'imaginary')
- The axiomatic definition of power $p = vi$ in which voltage and current are represented as complex numbers gives wrong *numerical* results; on the basis of Janet's heuristic rule, we have to multiply the voltage by the conjugate of the current (or vice-versa): $S = \dot{V}I^*$ i.e., $p = vi^{\dagger}$

○ *Vector representation*

Representing electrical magnitudes *as vectors* creates the following mathematical incongruities:

- Apparent power as a vector is a sum of a polar vector and an axial vector
- Active power, resulting from the inner product of two vectors, is a scalar; however, in the vector diagram, active power is represented as a vector
- Reactive power, resulting from the outer product of two polar vectors (voltage and current as polar vectors) is an axial vector. However, in the vector diagram it is represented as a polar vector

3. Conclusion

Each of the above mathematical systems used to represent electrical magnitudes gives rise to both mathematical ambiguities and physical misinterpretations. Not one of the mathematical formalisms is acceptable.

The current power paradigm is beset with epistemological and ontological flaws, such as the following:

- Although voltage and current are physically distinct, mathematically they are represented as identical
- The current paradigm corresponds mathematically to an Abelian gauge theory. The Abelian $U(1)$ symmetry (commutative symmetry) expresses the conservation of a physical entity, that is, the electric charge. This paradigm recognizes the electrical connectivity of the network; however, it ignores the magnetic linkages of the network.

I propose an electromagnetic power theory characterized by a higher symmetry $U(1) \times SU(2)$.

- The Abelian (commutative) symmetry takes into account the conservation of the electrical charge, whereas the non-Abelian (non-commutative) symmetry takes into account the conservation of the magnetic momentum

- Although electrons are carriers of electromagnetic matter, the power phenomena are perceived as merely electrical in nature. I conceive power and power phenomena as electromagnetic in nature

- The concept of instantaneous power transmission implies that we can observe and measure voltage and current simultaneously and with the same precision. This concept contradicts the theory of relativity, which postulates that the speed of physical processes cannot exceed the speed of light (c), as well as Heisenberg's uncertainty principle.

- The current paradigm neglects the time delay between cause and effect and fails to consider actions in the past and actions at a distance

- Voltage and current are considered as independent and separable physical entities, which they are not! Neither can exist without the other. They describe a single physical reality (energy-momentum) which has two different aspects

- Voltage and current should not be multiplied, but 'added' and treated as one mathematical entity (e.g. like the 'addition' of the real and imaginary parts of a complex number)

- Electrons are considered as carriers of electrical force, but they are not. Electrons are carriers of electromagnetic matter

- The role of photons as carriers of force (with light speed) and the existence of an additional field – the field of bosons – is ignored

- The process of transmutation (electromagnetic matter into electromagnetic energy and vice versa) is ignored

- The current paradigm ignores the gauge theory. In addition, it considers only Abelian symmetry, whereas transmission of energy-momentum at the subatomic scale also has a non-Abelian symmetry

In essence, the current power paradigm has not incorporated the last eighty years' achievements in quantum electrodynamics and restraint relativity. The current power paradigm is merely a calculation algorithm, a phenomenological description. It fails to give a causal explanation of the transmission process at the subatomic scale.

I dismiss not only the epistemic aspect of the current power paradigm (the Steinmetz-Janet symbolic method), but also the ontic interpretation of the concept of power and its 'mechanism' of transmission.

4. Hypothesis of a Quantum Electromagnetic Power Theory is Consistent with Quantum Electrodynamics and with the Theory of Restraint Relativity

The theory has an epistemic and an ontic part.

Epistemology of the theory

- Voltage and current are conceived as a unit: the 'voltampère' multivector $(v + Ii)$.
- Electromagnetic power is defined as the geometric product of the voltampère multivector and its complex conjugate. The result consists of two parts: 1) a scalar akin to active power representing the density of the potential electromagnetic energy, and 2) a trivector akin to reactive power representing the density of the kinetic electromagnetic energy and its electromagnetic momentum. Geometrically, the trivector can be visualized as a volume moving on a helicoidal trajectory.
- Electromagnetic power theory is endowed with a global (commutative) symmetry and a local (non-commutative) symmetry
- Electromagnetic power could be expressed in many other mathematical formalisms. I am using geometric algebra and the four-dimensional Minkowski space-time. However, I emphasize my support for mathematical pluralism

Ontology of the theory

- The theory is general, in the sense that it is applicable to DC and AC circuits
- Electromagnetic power is interpreted as a force akin to Faraday's electromagnetic force, represented in electrodynamics as the Riemann-Silberstein multivector
- In this theoretical hypothesis, at the mesoscopic level, electrons are interpreted as carriers of electromagnetic matter, whereas photons are carriers of electromagnetic force
- The non-Abelian mathematical symmetry is interpreted (on the basis of Noether's theorem) as reflecting the conservation of electromagnetic momentum
- The process of power transmission at the mesoscopic level is interpreted as a continuous chain: collision between electron and positron → annihilation of electromagnetic matter → transmutation into energy-momentum carried by photons with velocity of light → disintegration → transmutation of photons into an electron-positron pair, and so on. At the mesoscopic level, the carriers of electromagnetic energy-momentum (i.e. the photons) follow a helicoidal trajectory
- Although, investigation of the interaction between electromagnetic and weak forces is outside the scope of the monograph, the interpretation

of electromagnetic power as an electromagnetic force conforms with the view of quantum electrodynamics and particle physics
- At the macroscopic level, this new power theory bridges the gap between circuits carrying electrons and electromagnetic fields consisting of waves. At the mesoscopic level, it reveals the relationship between electrons and positrons (particles, leptons) and photons (waves, bosons)

5. Power Engineering Theory and Practice: Quo Vadis?

A rupture at theoretical level does not necessarily imply disruption of current engineering practice. I expect that, for economic, sociological, and psychological reasons, we will continue to use the current engineering procedures for analysis, design, operation, and control of power systems for some time to come.

The new theory, though not yet fully fledged, questions both the limits of our knowledge about power phenomena and our physical interpretation of those phenomena (i.e. the epistemology and ontology of power theory). The new theory is equivalent to a quantum leap at metalevel (in the sense of the philosophical interpretation of what constitutes a power theory)

Power theory, like any scientific theory, advances through conjectures and refutations. My conjectures must be submitted to theoretical discussions and experimental validation. I expect that this monograph will trigger a scientific debate with respect to the epistemology and ontology of power theory. This debate must be seen in the context of the larger debate regarding the Standard Model of physics.

Current power theory is still an unfinished, incomplete, and inconsistent theory. However, the ubiquitous and mundane-sounding power from the plug continues to be one of the most exciting domains of research. As Heaviside said, "There is no finality in a growing science," and power engineering is still a growing scientific discipline.

Research on power theory can be a very frustrating experience. Many times, while working on this monograph, I had feelings similar to those that Wolfgang Pauli expressed in a letter to his assistant:

"It is much too difficult for me and I wish that ... [I] had never heard anything of physics."(Pauli to Kronig, 21 May 1925).

However, I continued and will continue to work, taking solace in Hilbert's exhortation:

We must know – we will know![3]

---

[3] David Hilbert (1862-1945) – "Wir müssen wissen, wir werden wissen" - epitaph on Hilbert's tombstone in Göttingen and famous lines he spoke in his retirement address on 8 September, 1930

Bibliography

Abłamowicz, R. and G. Sobczyk (Eds.). (2004). *Lectures on Clifford (Geometric) Algebras and Applications*. New York: Springer Science+Business Media.

Abraham, M. (1898). *'Geometrische Grundbegriffe', Encyklopädie der mathematischen Wissenschaften mit Einschluss ihrer Anwendungen*, 3-47. Leipzig: Teubner.

Abraham, M. (1902). 'Prinzipien der Dynamik des Elektrons', *Annalen der Physik*, **315**(1): 105-179.

Abraham, M. (1932). *The Classical Theory of Electricity and Magnetism*. Rev. R. Becker, translated from the 8th German ed. (1930) by J. Dougall. London: Blackie.

Abur, A. and A.G. Exposito (2004). *Power System State Estimation: Theory and Implementation*. New York: Marcel Dekker.

Achard, F. (2005). 'James Clerk Maxwell, A treatise on electricity and magnetism', 1st ed. (1873). *In:* Landmark Writings in Western Mathematics 1640–1940. (Ed.) I. Grattan-Guinness, 564-587. Amsterdam: Elsevier Science.

Adler, C.G. (1976). 'Connection between conservation of energy and conservation of momentum', *American Journal of Physics*, **44**(5): 483-484.

Adler, R.B., LJ. Chu and R.M. Fano (1959). *Electromagnetic Energy Transmission and Radiation*. Cambridge: The MIT Press.

Aerts, D. (2009). 'Quantum particles as conceptual entities: A possible explanatory framework for quantum theory', *Foundations of Science*, **14**: 361-411.

Aguirre-Zamalloa, G., F.U. Arrue and J.R. Hernandez Gonzalez (2006). 'The three-phase instantaneous reactive power defined anew', *International Symposium on Power Electronics, Electrical Drives, Automation and Motion*, SPEEDAM, 2006: 1-6.

Ahlers, F.J. and U. Siegner (2011). 'The redefinition of the Ampere', *Proceedings of the 56th International Scientific Colloquium*, Ilmenau University of Technology, 12–16 September. https://nbn-resolving. org/urn:nbn:de:gbv:ilm1-2011iwk-151:3.

Aitchison, I.J.R. and A.J.G. Hey (2013). *Gauge Theories in Particle Physics*. Vol. 2: Non-Abelian Gauge Theories. Boca Raton, FL: CRC Press.

Akagi, H., Y. Kanazawa, K. Fujita and A. Nabae (1983). 'Generalized theory of instantaneous reactive power and its application', *Electrical Engineering in Japan*, **103**(4): 58-65.

Akagi, H., E.H. Watanabe and M. Aredes (2007). *Instantaneous Power Theory and Applications to Power Conditioning*. Piscataway, NJ: IEEE Press. https://doi.org/10.1002/9781119307181.

Akins, C.G. (2015). 'The electron as a confined photon'. *In:* The Nature of Light: What are Photons? VI, Proceedings of the Society of Photo-optical Instrumentation Engineers (SPIE), vol. 9570, (Eds). C. Roychoudhuri, A.F. Kracklauer and H. De Raedt, 957014. https://doi. org/10.1117/12.2187542.

Akins, C., R. Gauthier, A. Kracklauer, J. Macken, A. Meulenberg, *et al.* (2015). 'Are electrons oscillating photons, oscillating "vacuum", or something else? The 2015 panel discussion'. *In:* The Nature of Light: What are Photons? VI, Proceedings of the Society of Photo-optical Instrumentation Engineers (SPIE), vol. 9570, (Ed.). C. Roychoudhuri, A.F. Kracklauer and H. De Raedt, 957011. https://doi. org/10.1117/12.2205311.

Aleksandrov, A.D., A.N. Kolmogorov and M.A. Lavrent'ev (1963). *Mathematics, Its Contents, Methods, and Meaning*. Cambridge: The MIT Press.

Alger, P.L. and W.R. Oney (1954). Torque-energy relations in induction machines, *Transactions of the American Institute of Electrical Engineers, Part III: Power Apparatus and Systems*, **73**(2): 259-264.

Allen, L., M.W. Beijersbergen, R.J.C. Spreeuw and J.P. Woerdman (1992). 'Orbital angular momentum of light and the transformation of Laguerre-Gaussian laser modes', *Physical Review A*, **45**(11): 8185-8189.

Allen, L., M.J. Padgett and M. Babiker (1999). 'The orbital angular momentum of light'. *In:* Progress in Optics, vol. 39, (Ed.). Emil Wolf, 294-372. Amsterdam: Elsevier Science.

Allen, L. and M.J. Padgett (2002). 'Response to question #79. Does a plane wave carry spin angular momentum?' *American Journal of Physics*, **70**(6): 567-568.

Allen, L., M.J. Padgett and J. Courtial (2004). 'Light's orbital angular momentum', *Physics Today*, **57**(5): 35-40.

Aller, J.M., A. Bueno, J.A. Restrepo, M.I. Gimenez de Guzman and V.M. Guzman (1999). 'Advantages of the instantaneous reactive power definitions in three phase system measurement', *IEEE Power Engineering Review*, **19**(6): 54-56.

Aller, J.M., A. Bueno and T. Paga (2002). 'Power system analysis using space-vector transformation', *IEEE Transactions on Power Systems*, **17**(4): 957-965.

Altmann, S.L. (1986). *Rotations, Quaternions, and Double Groups*. Mineola, NY: Dover.

Altmann, S.L. (1989). 'Hamilton, Rodrigues, and the quaternion scandal',*Mathematics Magazine*, **62**(5): 291-308.

Altmann, S.L. and E.L. Ortiz (Eds.) (2005). 'Mathematics and social utopias in France: Olinde Rodrigues and his times', *History of Mathematics*, vol. 28. Providence, RI: American Mathematical Society.

Amin, B. (1994). 'Space phasor or space vector?' *European Transactions on Electrical Power*, **4**(1): 69-72.

Ampère, M. (1822). *Recueil d'observations– électrodynamiques*. Paris: Chez Crochard.

Anastasovski, P.K. et al. (2000). 'Classical electrodynamics without the Lorentz condition: Extracting energy from the vacuum', *PhysicaScripta*, **61**: 513-517.

Anderson, C.D. (1934). 'The positron', *Nature*, **133**: 313-316.

Anderson, J.W. (2005). *Hyperbolic Geometry*, 2nd ed., London: Springer.

Anderson, P.M. (1995). *Analysis of Faulted Power Systems*. New York: IEEE Press.

Anderson, R. and G.C. Joshi (1992). 'Quaternions and the heuristic role of mathematical structures in physics'. https://arxiv.org/abs/hep-ph/9208222v2.

Andreescu, T. and D. Andrica (2006). *Complex Numbers from A to...Z*. Boston: Birkhäuser.

Andronescu, Pl. (1935). 'Beitrag zum Problem der Wechselströme beliebiger Kurvenform', *Archiv für Elektrotechnik*, **29**: 802-806.

Andronescu, Pl. (1937a). 'Erwiderung', *Archiv für Elektrotechnik*, **31**(12): 833-834.

Andronescu, Pl. (1937b). 'Graphische Darstellung der Wirk-, Blind-, Verzerrungs- und Scheinleistung, so wie ein Beitrag zum Problem der Wechselströme beliebiger Kurvenform', *Archiv für Elektrotechnik*, **31**(3): 205-210.

Anglès, P. (2008). *Conformal Groups in Geometry and Spin Structures*. Boston: Birkhäuser.

Angot, A. (1957). *Compléments de mathématiques à l'usage des ingénieurs de l'électrotechnique et des télécommunications*. Paris: Éditions de la Revue d'Optique.

Antoniu, S. (1984). 'Le régime énergétique déformant: Une question de priorité', *Revue Générale de l'Électricité*, **93**(6): 357-362.

Appelquist, T., M.K. Gaillard and J.D. Jackson (1984). 'Physics at the superconducting super collider: A new high-energy accelerator would enable physicists to prove further into the basic constituents of matter and the forces that bind them together', *American Scientist*, **72**(2): 151-155.

Arabatzis, T. (2006). *Representing Electrons: A Biographical Approach to Theoretical Entities*. Chicago: University of Chicago Press.

Ariew, R. and P. Baker (1986). 'Duhem on Maxwell: A case-study in the interrelations of History of Science and Philosophy of Science', *Proceedings of the Biennial Meeting of the Philosophy of Science Association*, vol. 1, *Contributed Papers*, 145-156. Chicago: University of Chicago Press.

Arrayas, M., J.L. Trueba and A.F. Rañada (2012). 'Topological electromagnetism: Knots and quantization rules'. *In:* Trends in Electromagnetism: From Fundamentals to Applications, (Ed.). V. Barsan and R.P. Lungu, 71-88. IntechOpen. https://doi.org/10.5772/34703.

Arrillaga, J., C.P. Arnold and B.J. Harker (1980). *Computer Modelling of Electrical Power Systems*. Chichester, UK: Wiley.

Arsenovic, A. (2017). 'Applications of conformal geometric algebra to transmission line theory', *IEEE Access* **5**: 19920–19941. https://doi.org/10.1109/ACCESS.2017.2727819.

Arsenovic, A. (2019). 'A spinor model for cascading two port networks in conformal geometric algebra', Preprint, March 2019, ResearchGate. https://doi.org/10.13140/RG.2.2.17948.13445.

Arsenovic, A and A. Cortzen (2017). 'Using Cartan's rotation algorithm with conformal geometric algebra', Working Paper, July 2017, ResearchGate. https://doi.org/10.13140/RG.2.2.17721.88165.

Arthur, J.W. (2008). 'The fundamentals of electromagnetic theory revisited', *IEEE Antennas and Propagation Magazine*, **50**(1): 19-65.

Arthur, J.W. (2009). 'An elementary view of Maxwell's displacement current', *IEEE Antennas and Propagation Magazine*, **51**(6): 58-68.

Arthur, J.W. (2011). *Understanding Geometric Algebra for Electromagnetic Theory*, Piscataway, NJ: IEEE Press. https://doi.org/10.1002/9781118078549.

Arthur, J.W. (2013). 'The evolution of Maxwell's equations from 1862 to the present day', *IEEE Antennas and Propagation Magazine*, **55**(3): 61-81.

Assis, A.K.T. (1997). 'Circuit theory in Weber electrodynamics', *European Journal of Physics*, **18**: 241-245.

Assis, A.K.T. and J.A. Hernandes (2007). *The Electric Force of a Current: Weber and the Surface Charges of Resistive Conductors Carrying Steady Currents*. Montreal: Apeiron.

Assis, A.K.T. and H. Torres Silva (2000). 'Comparison between Weber's electrodynamics and classical electrodynamics', *Pramana – Journal of Physics*, **55**(3): 393-404.

Assis, A.K.T., J.A. Hernandes and J.E. Lamesa (2001). 'Surface charges in conductor plates carrying constant currents', *Foundations of Physics*, **31**(10): 1501-1511.

Atabekov, G.I. (1965). *Linear Network Theory*. Oxford: Pergamon Press.

Ayres, B. (1896). 'Annual meeting of the American Association for the Advancement of Science', *The Electrical World*, 5 September: 276-277.

Ayrton, W.E. and W.E. Sumpner (1890-1891). 'The measurement of the power given by any electric current to any circuit', *Proceedings of the Royal Society of London*, **49**: 424-439.

Ayrton, W.E. and W.E. Sumpner (1891). 'Alternate current and potential difference analogies in the methods of measuring power', *The London, Edinburgh and Dublin Philosophical Magazine and Journal of Science* (Ser. 5), **32**(195): 204-215.

Azzam, R.M.A. and N.M. Bashara (1977). *Ellipsometry and Polarized Light*. Amsterdam: North-Holland.

Babin, A. and A. Figotin (2016). *Neoclassical Theory of Electromagnetic Interactions: A Single Theory for Macroscopic and Microscopic Scales*. London: Springer.

Backhaus, U. and K. Schäfer (1986). 'On the uniqueness of the vector for energy flow density in electromagnetic fields', *American Journal of Physics*, **54**(3): 279-280.

Baez, J.C. (2002). 'The Octonions', *Bulletin of the American Mathematical Society*, **39**(2): 145-205.

Balci, M.E., M.H. Hocaoglu and S. Aksoy (2006). 'Transition from Poynting vector to instantaneous power', *Proceedings of the 12th International Conference on Harmonics & Quality of Power*, Cascais, Portugal. 5 p.

Baldomir, D. and P. Hammond (1996). *Geometry of Electromagnetic Systems*. Oxford: Clarendon Press.

Baldomir, D., M. Pereiro and J. Arias (2011). 'Geometrical information coded in Maxwell's equations: A review', *COMPEL: The International Journal for Computation and Mathematics in Electrical and Electronic Engineering*, **30**(2): 793-811.

Barany, M.J., A.-S. Paumier and J. Lützen (2017). 'From Nancy to Copenhagen to the world: The internationalization of Laurent Schwartz and his theory of distributions', *Historia Mathematica*, **44**(4): 367-394.

Bardham, D. and T.J. Osler (2002). 'An easy introduction to biplex numbers'. https://www.researchgate.net/publication/228812190.

Barnett, S.M. (2002). 'Optical angular-momentum flux', *Journal of Optics B: Quantum and Semi-classical Optics*, **4**: S7-S16.

Barnett, S.M. (2010). 'Resolution of the Abraham-Minkowski dilemma', *Physical Review Letters*, **104**: 1-4.

Barnett, S.M. and R. Loudon (2010). 'The enigma of optical momentum in a medium', *Philosophical Transactions of the Royal Society A: Mathematical, Physical, and Engineering Sciences*, **368**(1914): 927-939.

Barnett, S.M. and R. Zambrini (2007). 'Orbital angular momentum of light'. *In:* Quantum Imaging, (Ed.). M. Kolobov, 277-311. Springer.

Barret, J.-P., P. Bornard and B. Meyer (1997). *Power System Simulation.* London: Chapman & Hall.

Barrett, T.W. (1993). 'Electromagnetic phenomena not explained by Maxwell's equations', *In:* Essays on the Formal Aspects of Electromagnetic Theory, (Ed.). A. Lakhtakia, 6-86. Hackensack, New Jersey: World Scientific Publishing.

Barrett, T.W. (2000). 'Topology and the physical properties of electromagnetic fields', *Apeiron: A Journal for Ancient Philosophy and Science*, **7**(1-2): 3-11.

Barrett, T.W. (2001). *Topological Foundations of Electromagnetism, Annales de la Fondation Louis de Broglie*, **26** (numéro spécial): 55-79.

Barrett, T.W. (2008). *Topological Foundations of Electromagnetism*, World Scientific Series in Contemporary Chemical Physics, 26. World Scientific Publishing Co.

Bateman, H. (1915). *The Mathematical Analysis of Electrical and Optical Wave Motion on the Basis of Maxwell's Equations.* Cambridge: Cambridge University Press.

Bateman, H. (1932-1944). *Partial Differential Equations of Mathematical Physics.* Cambridge: Cambridge University Press.

Batterman, R.W. (2014). 'The inconsistency of Physics (with capital 'P')', *Synthese*, **191**(13): 2973-2992.

Baylis, W.E. (1996). *Clifford (Geometric) Algebra with Applications in Physics, Mathematics, and Engineering.* Boston: Birkhäuser.

Baylis, W.E. (1999). *Electrodynamics: A Modern Geometric Approach.* Boston: Birkhäuser.

Baylis, W.E. (2004). 'Geometry of paravector space with applications to relativistic physics'. *In:* Computational Non-commutative Algebra and Applications, (Ed.). J. Byrnes, 363-387. Kluwer Academic Publishers.

Baylis, W.E. and G. Sobczyk (2004). 'Relativity in Clifford's geometric algebras of space and space-time', *International Journal of Theoretical Physics*, **43**(10): 2061-2079.

Bayro-Corrochano, E. (2010). *Geometric Computing: For Wavelet Transforms, Robot Vision, Learning, Control and Action.* London: Springer.

Bearden, T.E. (2001). 'Energy from active vacuum: The motionless electromagnetic generator'. *In:* Modern Nonlinear Optics, vol. 2, (Ed.). M.E. Evans, 699-776, Wiley.

Beaty, H.W. (2001). *Handbook of Electric Power Calculations.* New York: McGraw-Hill.

Beck, Y., N. Calamaro and D. Shmilovicz (2016). 'A review study of instantaneous electric energy transport theories and their novel implementations', *Renewable and Sustainable Energy Reviews*, **57**: 1428-1439.

Becker, R. (1944). *Theorie der Elektrizität*, vol. 1: *Einführung in die Maxwellsche Theorie der Elektrizität*. Leipzig: Teubner.

Becker, R. (1949). *Theorie der Elektrizität*, vol. 2: *Elektronentheorie*. Leipzig: Teubner.

Becker, R. and F. Sauter (1968). *Theorie der Elektrizität*, vol. 3: *Elektrodynamik der Materie*. Stuttgart: Teubner.

Bedell, F. (1896a). 'Admittance and Impedance loci', *Physical Review* (Series 1), **4**: 143-149.

Bedell, F. (1896b). 'XXXII. Admittance and Impedance loci', *The London, Edinburgh, and Dublin Philosophical Magazine and Journal of Science* (5th Series), **42**(257): 300-308.

Bedell, F. (1906). 'Note on the graphical representation of non-sinusoidal alternating currents', *Physical Review*, 23: 249.

Bedell, F. (1927). 'Non-harmonic alternating currents', *Transactions of the American Institute of Electrical Engineers*, 46: 648-659.

Bedell, F. and A.C. Crehore (1892). 'Propagation of the current in a cable containing distributed static capacity and self-induction', *The Electrician*, **29**(751): 619-621.

Bedell, F. and A.C. Crehore (1893a). 'Note: Geometrical proof of the three-ammeter method of measuring power', *Physical Review* (Series 1), 1: 59-61.

Bedell, F. and A.C. Crehore (1893b). 'General discussion of the current flow in two mutually related circuits containing capacity', *Physical Review* (Series 1), 1: 117-126.

Bedell, F. and A.C. Crehore (1895). *Alternating Currents. An Analytical and Graphical Treatment for Students and Engineers*, 3rd ed. New York: W.J. Johnston Co.

Bedell, F. and A.C. Crehore (1901). *Alternating Currents: An Analytical and Graphical Treatment for Students and Engineers*. Ithaca, NY: Electrical World and Engineer.

Bedell, F. and E.C. Mayer (1915). 'Distortion of alternating-current wave caused by cyclic variation in resistance', presented at the *3rd Midwinter Convention of the American Institute of Electrical Engineers*, New York, February 18: 333-342.

Belevitch, V. (1962). 'Summary of the history of circuit theory', *Proceedings of the IRE* [Institute of Radio Engineers], **50**(5): 848-855.

Bell, J.S. (2004). *Speakable and Unspeakable in Quantum Mechanics*. Cambridge: Cambridge University Press.

Beller, M. (1999). *Quantum Dialogue. The Making of a Revolution*. Chicago: The University of Chicago Press.

Belloni, L. (1981). 'Historical remarks on the 'classical' electron radius', *Lettere al Nuovo Cimento*, 31: 131-134.

Belot, G. (1998). 'Understanding electromagnetism',*The British Journal for the Philosophy of Science*, **49**(4): 531-555.

Benischke, G. (1922). *Die wissenschaftlichen Grundlagen der Elektrotechnik*. Berlin: Verlag von Julius Springer.

Benn, I.M. and R.W. Tucker (1987). *An Introduction to Spinors and Geometry with Applications in Physics*. Bristol: Adam Hilger.

Berg, E.J. (1907). *Electrical Energy: Its Generation, Transmission, and Utilization: Lectures given at Union University*. New York: McGraw-Hill.

Berg, E.J. (1911). 'The direction of rotation in alternating-current vector diagrams', a paper presented at the Pittsfield-Schenectady Mid-year convention of the American Institute of Electrical Engineers, February 16: 575-579.

Berg, E.J. (1924). 'Heaviside's operators in engineering and physics. Lectures at the Moore School of Electrical Engineering, February-March', *Journal of the Franklin Institute*, 198: 647.

Berg, E.J. (1929). *Heaviside's Operational Calculus as Applied to Engineering and Physics*. New York: McGraw-Hill.

Berg, L. (1965). *Einführung in die Operatorenrechnung*. Berlin: *VEB Deutscher Verlag der Wissenschaften*.

Bergen, A.R. and V. Vittal (1986). *Power System Analysis*. Englewood Cliffs: Prentice-Hall, Inc.

Bergman, D.L. and J.P. Wesley (1990). 'Spinning charged ring model of electron yielding anomalous magnetic moment', *Galilean Electrodynamics*, **1**(5): 63-67.

Besso, M.A. (1900). 'Symbolische Darstellung doppelperiodischer Vektorprodukte und allgemeiner Wechselstromwellen', *Elektrotechnische Zeitschrift*.

Bessonov, E.G. (1997). 'The foundations of classical electrodynamics'. arXiv: physics/9708002v2 [physics.class-ph] 21 Aug.

Beth, R.A. (1936). 'Mechanical detection and measurement of the angular momentum of light', *Physical Review* 50: 115-125.

Bevelacqua, J.J. (2012). 'Physical interpretation of electrodynamics within the Yang-Mills theory', *Physics Essays*, **25**(3): 369-373.

Bialynicki-Birula, I. and Z. Bialynicka-Birula (2006). 'Beams of electromagnetic radiation carrying angular momentum: The Riemann-Silberstein vector and the classical-quantum correspondence', *Optics Communications*, 264: 342-351.

Bialynicki-Birula, I. and Z. Bialynicka-Birula (2012). 'The role of the Riemann-Silberstein vector in classical and quantum theories of electromagnetism'. arXiv:1211.2655v2 [physics.class-ph] 28 December 2012.

Bialynicki-Birula, I. and Z. Bialynicka-Birula (2013). 'The role of the Riemann-Silberstein vector in classical and quantum theories of electromagnetism', *Journal of Physics A: Mathematical and Theoretical*, **46**(5): 053001.

Bialynicki-Birula, I. and Z. Bialynicka-Birula (2013). 'Corrigendum: The role of the Riemann-Silberstein vector in classical and quantum theories of electromagnetism', *Journal of Physics A: Mathematical and Theoretical*, **46**(15): 159501.

Bissell, C. (2012). 'Metatools for information engineering design'. In: *Ways of Thinking, Ways of Seeing*, (Ed.). C. Bissell and C. Dillon, 71-94. Springer.

Blakesley, T.H. (1887). 'On a method of determining the difference between the phase of two harmonic currents of electricity having the same period', *Proceedings of The Physical Society of London*, 9:165-167 (*read* March 10, 1888).

Blakesley, T.H. (1889). *Papers on Alternating Currents of Electricity, for the Use of Students and Engineers*. London: Whittaker & Co and George Bell & Sons.

Blakesley, T.H. (1891). 'XXXIX. Further contributions to dynamometry, or the measurement of power', *The London, Edinburgh, and Dublin Philosophical Magazine and Journal of Science*, Series 5, **31**(191): 346-354.

Blakesley, T.H. (1892). 'A possible misunderstanding', *Nature*, **45**(1167): 441.

Blinder, S.M. (2003). 'Singularity-free electrodynamics for point charges and dipoles: Classical model for electron self-energy and spin', *European Journal of Physics*, **24**(3): 271-275.

Bliokh, K.Y. and F. Nori (2011). 'Characterizing optical chirality', *Physical Review* A, 83.021803: 3 p.

Bliokh, K.Y. and F. Nori (2015). 'Transverse and longitudinal angular momentum of light', *Physics Reports*. 592: 1-38.

Bloch, F. (1929). 'Über die Quantenmechanik der Elektronen in Kristallgittern', *Zeitschrift für Physik*, 52: 555-600.

Bloch, F. and A. Nordsieck (1937). 'Note on the radiation field of the electron', *Physical Review*, **52**(2): 54-59.

Bochner, S. (1963). 'The significance of some basic mathematical conceptions for physics', *Isis*, **54**(2): 179-205.

Bohm, D. (1951). *Quantum Theory*. Englewood Cliffs, NJ: Prentice-Hall.

Bohr, Niels (1913). 'On the constitution of atoms and molecules', *The London, Edinburgh, and Dublin Philosophical Magazine and Journal of Science*, Series 6, **26**(151): 1-25.

Boldt, E. (1965). 'The Poynting vector in magnetohydrodynamics', *American Journal of Physics*, **33**(4): 298-300.

Bolinder, E.F. (1987). 'Clifford Algebra: What is it?' *IEEE Antennas and Propagation Society Newsletter*, **29**(4): 18-23.

Bolton, W. (1995). *Complex Numbers*. Longman Scientific & Technical.

Bolund, B., M. Leijon and U. Lundin (2008). 'Poynting theorem applied to cable wound generators' *IEEE Transactions on Dielectrics and Electrical Insulation*, **15**(2): 600-605.

Bork, A.M. (1966). 'Vectors versus Quaternions: The letters in *Nature*', *American Journal of Physics*, **34**(3): 202-211.

Bork, A.M. (1966). 'Physics just before Einstein', *Science*, **152**(3722): 597-603.

Bork, A.M. (1967a). 'Maxwell and the electromagnetic wave equation', *American Journal of Physics*, **35**(9): 844-849.

Bork, A.M. (1967b). 'Maxwell and the vector potential', *Isis*, **58**(2): 210-222.

Bork, A.M. (1968). 'Quaternions and other topics', Review of *The Mathematical Papers of Sir William Rowan Hamilton*, vol. 3, *Algebra*, by Sir William Rowan Hamilton, H. Halberstam and R.E. Ingram, *Science*, **160**(3828): 663-664.

Born, M. (1969). *Physics in My Generation*. New York: Springer Science.

Born, M. and E. Wolf (1959). *Principles of Optics: Electromagnetic Theory of Propagation, Interference and Diffraction of Light*. Pergamon Press, London.

Bostick, W.H. (1986). 'What laboratory-produced plasma structures can contribute to the understanding of cosmic structures both large and small', *IEEE Transactions on Plasma Science*, PS, **14**(6): 703-717.

Bourgin, D.G. and R.J. Duffin (1937). 'The Heaviside Operational Calculus', *American Journal of Mathematics*, **59**(3): 489-505.

Bowen, M. and J. Coster (1980) . 'Born's discovery of the quantum-mechanical matrix calculus', *American Journal of Physics*, **48**(6): 491-492.

Bracco, C. (2014). 'Einstein and Besso: From Zürich to Milan' *Istituto Lombardo – Accademia di Scienze e Lettere Rendiconti di Scienze*, 148: 285-322. https://doi.org/10.4081/scie.2014.178.

Bravo, J.C. (2008). *Representación multivectorial de la Potencia Aparente en regímenes periódicos n-senoidales aplicando algebras de Clifford*, doctoral dissertation, University of Seville, Spain.

Bravo, J.C., M. Castilla, J.C. Montaño, M. Ordoñez, M.V. Castilla, A. Lopez, D. Borras and J. Gutierrez (2010). 'Non-active power multivector, 15th IEEE Mediterranean Electrotechnical Conference (MELECON)', Malta, 25-28 April 2010, *IEEE*. 9978-1-4244-5795-3/10:1021-1026.

Breisig, F. (1900). 'Über die graphische Darstellung des Verlaufes von Wechselstromes längs langer Leitungen', *Elektrotechnische Zeitschrift* (ETZ), **25**(4): 87-92.

Breisig, F. (1924). *Theoretische Telegraphie: Eine Anwendung der Maxwellschen Elektrodynamik auf Vorgänge in Leitungen und Schaltungen*. Wiesbaden: Springer Fachmedien Wiesbaden GmbH.

Breitfeld, C. (1912). *Berechnung von Wechselstrom-Fernleitungen*. Braunschweig: Druck und Verlag von Fiedr. Vieweg & Sohn.

Brychkov, Yu. A., A.P. Prudnikov and V.S. Shishov (1981). *Operational Calculus*, translated from *Itogi Naukii Tekniki, Serya Matematicheskii Analiz*, 16: 99-148, *Journal of Soviet Mathematics*, 15: 6(1981), 733-765.

Brill, M.H., D. Dameron and T.E. Phipps (2011). 'A misapprehension concerning electric current neutrality', *Physics Essays*, **24**(3): 325-326.

Brittain, J.E. (1972). Review of *Lectures on Electrical Engineering* by Charles Proteus Steinmetz, (Ed.). Philip L. Alger, *Isis*, **63**(2): 299-300.

Bromwich, T. J. l'A. (1928a). 'LXXXIV. Heaviside's formulae for alternating currents in cylindrical wires', *The London, Edinburgh, and Dublin Philosophical Magazine and Journal of Science* Series 7, **6**(38): 842-854.

Bromwich, T. J. l'A. (1928b). 'Note on Prof. Carslaw's Paper', *The Mathematical Gazette*, **14**(196): 226-228.

Brown, H.E. (1985). *Solution of Large Networks by Matrix Methods*, 2nd ed. New York: Wiley.

Buchwald, J.Z. (1985). *From Maxwell to Microphysics. Aspects of Electromagnetic Theory in the Last Quarter of the Nineteenth Century*. Chicago: University of Chicago Press.

Buchwald, J.Z. and A. Warwick (Eds.) (2001). *Histories of the Electron: The Birth of Microphysics*, Dibner Institute Studies in the History of Science and Technology. Cambridge: MIT Press.

Budeanu, C. (1927). *Puissances réactives et fictives, Institut National Roumain pour l'étude de l'aménagement et de l'utilisation des sources d'énergie 2*. Bucharest: *Impr. Cultura Naţională*.

Budeanu, C. (1937). 'Graphische Darstellung der Wirk-, Blind-, Verzerrungs- und Scheinleistung, sowie ein Beitrag zum Problem der Wechselströme beliebiger Kurvenform', *Archiv für Elektrotechnik*, **31**(12): 832-833.

Budeanu, C. (1938). 'Kapazitäten und Induktivitätenals Verzerrende Elemente', *Archiv für Elektrotechnik*, **32**(4): 251-259.

Bueno, M.A. and A.K.T. Assis (2001). *Inductance and Force Calculations in Electrical Circuits*. Huntington, NY: Nova Science Publishers.

Bunet, P. (1926). 'Puissance réactive et harmoniques', *Revue Générale de l'Électricité*, **19**(10): 365-372.

Bunge, M. (1950). 'The inexhaustible electron', *Science & Society*, **14**(2): 115-121.

Bunge, M. (1967). *Foundations of Physics*. New York: Springer.

Bunge, M. (1973). *Philosophy of Physics*. Dordrecht: D. Reidel.

Bunge, M. (2000). 'Energy: Between Physics and Metaphysics', *Science and Education*. 9: 457-461.

Bureau International des Poids et Mesures (2019). *The International System of Units (SI)*, 9th ed., vol. 1.06, 216 p.

Burkhardt, H. (1908). 'Entwicklungen nach oscillirenden Funktionen und Integration der Differential-gleichungen der mathematischen Physik', *Jahresbericht der Deutschen Mathematiker-Vereinigung*, **10**(2): 823 ff.

Bush, V. (1929). *Operational Circuit Analysis*. New York: Wiley.

Byk, A. (1930a). 'Komplexe und ebene Vektorrechnung in der Wechselstrom-technik'. In: *Forschung und Technik*, (Ed.). W. Petersen, 84-103. *Allgemeine Elektrizitäts-Gesellschaft, Verlag von Julius Springer*.

Byk, A. (1930b). 'Kovariante Tensorformen des Ohmschen und des Jouleschen Gesetzes', *Zeitschrift für Physik A Hadrons and Nuclei*, **65**(7-8): 517-540.

Cable, J. (1964). 'What is a vector?' *The Mathematical Gazette*, **48**(363): 34-36.

Cahen, F. (1962). *Électrotechnique, Tome 1, Circuits et réseaux en régime permanent*. Paris: Gauthier-Villars.

Cakareski, Z. and A.E. Emanuel (1999). 'On the physical meaning of non-active powers in three-phase systems', *IEEE Power Engineering Review*, **19**(7): 46-47.

Cakareski, Z. and A.E. Emanuel (2001). 'Poynting vector and the quality of electric energy', *European Transactions on Electrical Power*, **11**(6): 375-381.

Calamaro, N., Y. Beck and D. Shmilovitz (2015). 'A review and insights on Poynting vector theory and periodic averaged electric energy transport theories', *Renewable and Sustainable Energy Reviews*, 42: 1279-1289.

Calvaer, A. (1956). *Théorie des réseaux en quadrature antisymétriquement couplés et Applications à L'Électrotechnique, Thèse d'Agrégation de l'Enseignement Supérieur*. Liège: Extrait du Bulletin Scientifique de l'A.I.M.

Cambier, J-L, D.A. Micheletti (2000). 'Theoretical analysis of the electron spiral toroid concept', NASA/CR-2000-210654 (December 2000), 44 p.

Cameron, R.P. and S.M. Barnett (2012). 'Electric-magnetic symmetry and Noether's theorem', *New Journal of Physics*, 14: 123019 (27 p). https://doi.org/10.1088/1367-2630/14/12/123019.

Campos, I. and J.L. Jiménez (1992). 'About Poynting's theorem', *European Journal of Physics*, **13**(3): 117-121.

Carroll, J.E. (2003). 'Traditional vectors as an introduction to geometric algebra', *European Journal of Physics*, **24**(4): 419-427.

Carslaw, H.S. (1928). 'Operational Methods in Mathematical Physics', *The Mathematical Gazette*, **14**(196): 216-225.

Carson, J.R. (1927). 'Electromagnetic theory and the foundations of electric circuit theory', *The Bell System Technical Journal*, **6**(1): 1-17.

Carson, J.R. (1930). 'Notes on the Heaviside operational calculus', *The Bell System Technical Journal*, **9**(1): 150-162.

Case, M.J. and P.H. Swart (1996). 'The Gabor transform as a power-system analysis tool', *European Transactions on Electrical Power*, **6**(6): 387-390.

Casper, L. (1926). 'Die Operatorenrechnung', *Archiv für Elektrotechnik*, 16: 267-272; 367-369.

Castilla, M.V. (2013). 'Control of disturbing loads in residential and commercial buildings via geometric algebra', *The Scientific World Journal*, 2013: 463983. https://doi.org/10.1155/2013/463983.

Castilla, M. and J.C. Bravo (2009). 'An approach to the multivectorial apparent power in terms of a generalized Poynting multivector', *Progress in Electromagnetics Research B*, 15:401-422.

Castilla, M., J.C. Bravo, M. Ordoñez and J.C. Montaño (2008a). 'Clifford theory: A geometrical interpretation of multivectorial apparent power', *IEEE Transactions on Circuits and Systems I: Regular papers*, **55**(10): 3358-3367.

Castilla, M., J.C. Bravo and M. Ordoñez (2008b). 'Geometric Algebra: A multivectorial proof of Tellegen's theorem in multiterminal networks', *IET Circuits, Devices & Systems*, **2**(4): 383-390.

Castilla, M., Juan Carlos Bravo, Manuel Ordoñez, Juan-Carlos Montaño, A. Lopez, D. Borras and J. Gutierrez (2008c). 'The geometric algebra as a power theory analysis tool', Conference Paper. International School on Nonsinusoidal Currents and Compensation, Lagow, Poland, *Przeglad Elektrotechniczny*, **85**(1):1-7. doi 10.1109/ISNCC.2008.4627490.

Castilla, M., J.C. Bravo, M. Ordoñez and J.C. Montaño (2010). 'The role of Poynting vector in power theory', Conference Paper, International School on Non-sinusoidal Currents and Compensation, Lagow, Poland, June 15-18, 4p.. https://doi.org/10.1109/isncc.2010.5524519.

Castilla, M., J.C. Bravo, J.C. Montaño, M. Ordoñez and A. Lopez (2011). 'Considerations on the non-active power using geometric algebra', *Proceedings of the 2011 International Conference on Power Engineering, Energy and Electrical Drives*, Torremolinos, Spain, 4 p. 978-1-4244-9843-7/1.

Castilla, M.V., J.C. Bravo and F.I. Martin (2018). 'Multivectorial strategy to interpret a resistive behavior of loads in smart buildings', *IEEE 12th International Conference on Compatibility, Power Electronics and Power Engineering* (CPE-POWERENG), Doha, Qatar, 10-12 April, 2018. 5 p.

Castro-Nuñez, M.D. (2013). *The Use of Geometric Algebra in the Analysis of Non-sinusoidal Networks and Construction of a Unified Power Theory for Single-phase Systems: A Paradigm Shift*, Ph.D. Thesis, University of Calgary, Canada.

Castro-Nuñez, M.D. and R. Castro-Puche (2012a). 'Advantages of geometric algebra over the complex numbers in the analysis of networks with non-sinusoidal sources and linear loads', *IEEE Transactions on Circuits and Systems – I: Regular Papers*, **59**(9): 2056-2064.

Castro-Nuñez, M. and R. Castro-Puche (2012b). 'The IEEE Standard 1459, the CPC power theory, and Geometric Algebra in circuits with non-sinusoidal sources and linear loads', *IEEE Transactions on Circuits and Systems – I: Regular Papers*, **59**(12): 2980-2990.

Castro-Nuñez, M.D., R. Castro-Puche and E. Nowicki (2010). The use of geometric algebra in circuit analysis and its impact on the definition of power, *Proceedings of the 2010 International School on Non-sinusoidal Currents and Compensation* (ISNCC), Lagow, Poland, 89-95.

Castro-Nuñez, M.D., D. Londoño-Monsalve and R. Castro-Puche (2016). 'M, the conservative power quantity based on the flow of energy', *Journal of Engineering*, **2016**(7): 269-276.

Catoni, F., *et al.* (2004). 'Two-dimensional hypercomplex numbers and related trigonometries and geometries', *Advances in Applied Clifford Algebras*, **14**(1): 47-68.

Cauer, W. (1954). *Theorie der linearen Wechselstrom-Schaltungen*. Berlin: Akademie-Verlag.

Chabay, R. and B. Sherwood (1995). *Electric and Magnetic Interactions*. New York: Wiley.

Chabay, R. and B. Sherwood (2006). 'Restructuring the introductory electricity and magnetism course', *American Journal of Physics*, **74**(4): 329-336.

Chadwick, J., P. Blackett and G. Occhialini (1933). 'New Evidence for the Positive Electron', *Nature* 131: 473.

Chalmers, A.F. (1973a). 'Maxwell's methodology and his application of it to electromagnetism', *Studies in History and Philosophy of Science*, **4**(2): 107-164.

Chalmers, A.F. (1973b). 'The limitations of Maxwell's electromagnetic theory', *Isis*, **64**(4): 469-463.

Chalmers, A.F. (1975). ''Maxwell and the displacement current', *Physics Education*, **10**(1): 45-49.

Chalmers, A.F. (2013). *What is This Thing Called Science?* Indianapolis, Cambridge: Hackett.

Chappell, J.M., A. Iqbal and D. Abbott (2010). 'A simplified approach to electromagnetism using geometric algebra'. arXiv:1010.4947v2 [physics. Ed.-ph] 10 Nov.: 1-30.

Chappell, J.M., A. Iqbal, N. Ianella and D. Abbott (2012). 'Revisiting special relativity: A natural algebraic alternative to Minkowski space-time', *PLOS One*, **7**(12): 1-10.

Chappell, J.M., S.P. Drake, C.L. Seidel, L.J. Gunn, A. Iqbal, A. Allison and D. Abbott (2014). 'Geometric algebra for electrical and electronic engineers', *Proceedings of the IEEE*, **102**(9): 1340-1363.

Chappell, J.M., A. Iqbal and D. Abbott (2015). 'Geometric algebra: A natural representation of three-space', 128 p. arXiv:1101.3619v3 [physics. hist-ph] 16 Apr. 2015.

Chappell, J.M., A. Iqbal, J.G. Hartnett and D. Abbott (2016). 'The vector algebra: A historical perspective', *IEEE Access*, 4: 1997-2004.

Chatwin, P.C. (1984). 'Review of Operational Calculus', Vol. 1, by Jan Mikusinski, *The Mathematical Gazette*, **68**(446): 310-311.

Chaves-Jiménez, A. and D. Jeltsema (2011). 'Application of geometric algebra for power factor improvement: A survey', Poster presented at the Conference on Technologies for Sustainable Development TDS2011, Cartago, Costa Rica, February 3-4, 2011.

Chaves-Jiménez, A., D. Jeltsema and J. van der Woude (2011). 'Application of geometric algebra for power factor improvement;, *Proceedings of the 33rd International Telecommunications Energy Conference (INTELEC).* Amsterdam, Netherlands, October 2011, 1-6.

Chubykalo, A., A. Espinoza and R. Tzonchev (2004). 'Experimental test of compatibility of the definitions of the electromagnetic energy density and the Poynting vector', *The European Physical Journal D - Atomic, Molecular, Optical and Plasma Physics*, 31: 113-120.

Ciuceanu, R.M., I.V. Nemoianu, V. Manescu (Paltanea), G. Paltanea (2018). 'On professor Ţugulea's visionary power theory: A review, recent advances and perspectives', *Revue Roumaine des Sciences Techniques-Series Électrotechnique et Énergétique*, 63(2): 123-127.

Clarke, E. (1943). *Circuit Analysis of A-C Power Systems*. New York: Wiley.

Clifford, W.K. (1878). 'Applications of Grassmann's Extensive Algebra', *American Journal of Mathematics*, 1(4): 350-358.

Clifford, W.K. (1882). *Mathematical Papers*, (Ed.). Robert Tucker. London: MacMillan.

Coburn, N. (1955). *Vector and Tensor Analysis*. New York: MacMillan.

Coelho, R.L. (2009). 'On the concept of energy: History and philosophy for science teaching', *Procedia Social and Behavioral Sciences*, 1(1): 2648-2652.

Coelho, R.L. (2014). 'On the concept of energy: Eclecticism and Rationality', *Science & Education*, 23(6): 1361-1380.

Cohen, I.B. (1987). 'Faraday and Franklin's 'Newborn Baby'', *Proceedings of the American Philosophical Society*, 131(2): 177-182.

Cohen, J., F. de Leon and L.M. Hernandez (1998). 'Physical time domain representation of powers in linear and nonlinear electrical circuits', *IEEE Transactions on Power Delivery*, 14(4): 1240-1246.

Cole, D.C. (1999). 'Cross-term conservation relationship for electromagnetic energy, linear momentum, and angular momentum', *Foundations of Physics*, 29(11): 1673-1693.

Collin, R.E. (1998). 'Minimum Q of small antennas', *Journal of Electromagnetic Waves and Applications*, 12(10): 1369-1393.

Colyvan, M. (2008). 'The ontological commitments of inconsistent theories', *Philosophical Studies*, 141(1): 115-123.

Consa, O. (2014). 'Helical model of the electron', *The General Science Journal* (June 10, 2014), 14 p.

Consa, O. (2017). 'Electron toroidal moment', *The General Science Journal* (December 29, 2017), 6 p.

Consa, O. (2018). 'Helical solenoid model of the electron', *Progress in Physics*, 14(2): 80-89.

Conway, J.H. and D.A. Smith (1970). *On quaternions and octonions: Their Geometry, Arithmetic and Symmetry*. Natick, MA: A.K. Peters.

Cooper, J.L.B. (1952). 'Heaviside and the Operational Calculus', *The Mathematical Gazette*, **36**(315): 5-19.

Coopersmith, J. (2010). *Energy, the Subtle Concept: The Discovery of Feynman's Blocks from Leibniz to Einstein*. Oxford University Press.

Cornille, P. (2001). 'Electrodynamics and topology'. In: *Modern Nonlinear Optics*, Part 3, 2nd ed., (Ed.). M.W. Evans, 557-610, *Advances in Chemical Physics*, 119. John Wiley & Sons.

Cote, P.J. (2009). 'New perspectives on classical electromagnetism', arXiv: 0903.4104v4 [physics.class-ph] 2011. Pp. 1-16.

Cote, P.J. and M.A. Johnson (2008). 'Required revisions to classical electromagnetism', *Technical Report AREAW-TR-08014* (U.S. Army): 1-17.

Cote, P.J. and M.A. Johnson (2010). Comments on 'What the vector potential measures' by E.J. Konopinski, *Technical Report ARWSB-TR-11012* (U.S. Army): 1-4.

Cote, P.J., M. Johnson and S.L. Makowiec (2011). 'Groupthink and the blunder of the gauges', *Technical Report ARWSB-TR-11038* (U.S. Army): 1-11.

Craig, V. (1951). 'Vector Analysis', *Mathematics Magazine*, **25**(2): 67-86.

Crainic, E.D. and A.I. Petroianu (2007a). 'An affine geometrical approach to power systems problems'. In: *Power Plants and Power Systems Control*, (Ed.). D. Westwick, 101-106. Oxford: Elsevier IFAC Publications.

Crainic, E.D. and A.I. Petroianu (2007b). 'Application of affine transformations to real-time power system EMS functions'. In: *Power Plants and Power Systems Control*, (Ed.). D. Westwick, 333-338. Oxford: Elsevier IFAC Publications.

Crastan, V. (2012). *Elektrische Energieversorgung*. Berlin: Springer.

Crenshaw, M.E. (2012). 'Electromagnetic momentum in a dielectric and the energy-momentum tensor', arXiv: 1211.1935v1 [physics.class-ph] (8 Nov. 2012): 1-8.

Crenshaw, M.E. (2014). 'Electromagnetic momentum and the energy-momentum tensor in a linear medium with magnetic and dielectric properties', *Journal of Mathematical Physics*, **55**(4): 1-12.

Crilly, T. (2003). 'An Argand diagram for two by two matrices', *The Mathematical Gazette*, **87**(509): 209-216.

Cristaldi, L. and A. Ferrero (1996). 'Mathematical foundations of the instantaneous power concepts: An algebraic approach', *European Transactions on Electrical Power*, **6**(5): 305-309.

Crow, M. (2003). *Computational Methods for Electric Power Systems*. Boca Raton, FL: CRC Press.

Crowe, M.J. (1975). 'Ten 'laws' concerning patterns of change in the history of mathematics', *Historia Mathematica*, **2**(2): 161-166.

Crowe, M.J. (1967; 1985) (1994). *A History of Vector Analysis: The Evolution of the Idea of a Vectorial System*. New York: Dover.

Czarnecki, L.S. (1983). 'An orthogonal decomposition of the current of non-sinusoidal voltage sources applied to non-linear loads', *International Journal of Circuit Theory and Applications*, 11(2): 235-239.

Czarnecki, L.S. (1996). 'Discussion D3', *European Transactions on Electrical Power*, 6(5): 313-314.

Czarnecki, L.S. (1997). 'Budeanu and Fryze: Two frameworks for interpreting power properties of circuits with non-sinusoidal voltages and currents',*Electrical Engineering*, 80(6): 359-367.

Czarnecki, L.S. (1999). 'Harmonics and power phenomena', *Wiley Encyclopedia of Electrical and Electronic Engineering*. New York: Wiley: 1-27.

Czarnecki, L.S. (2000). 'Energy flow and power phenomena in electrical circuits: Illusions and reality', *Electrical Engineering*, 82(3-4): 119-126.

Czarnecki, L.S. (2004). 'Considerations on the concept of Poynting vector contribution to power theory development', paper read at the Sixth International Workshop on Power Definitions and Measurements under Non-sinusoidal Conditions, Milano, Italy, October 2003, *L'Energia Elettrica*, 81: 64-74.

Czarnecki, L.S. (2007). 'Closure on 'Could power properties of three-phase systems be described in terms of Poynting vector?',*IEEE Transactions on Power Delivery*, 22(2): 1269-1270.

Czarnecki, L.S. (2008). 'Currents' Physical Components (CPC) concept: A fundamental of power theory', International School on Non-sinusoidal Currents and Compensation, Lagow, Poland 2008, IEEE, 978-1-4244-2130; *Przeglad Elektrotechniczny*, 84(6): 1-11.

Czarnecki, L.S. (2013). 'Meta-theory of electric powers and present state of power theory with periodic voltages and currents', *Przeglad Elektrotechniczny*, 89(6): 26-31.

Czarnecki, L.S. (2016). 'From Steinmetz to currents' physical components (CPC): History of power theory development', *IEEE International Conference on Applied and Theoretical Electricity* (CATE), Craiova, Romania, 978-1-4673-8562: 6-16.

Czarnecki, L.S. (2020). 'Do energy oscillations degrade the energy transfer in electrical systems?' 2020, *IEEE Texas Power and Energy Conference* (TPEC), 6p, College Station, TX. 978-1-7281-4436-8/20. .

D'Agostino, S. (1975). 'Hertz's researches on electromagnetic waves', *Historical Studies in the Physical Sciences*, 6: 261-323.

D'Agostino, S., S. Leva and A.P. Morando (2000). '*Per una storia della teoria delle reti elettriche*', *Atti del XIX Congresso Nazionale di Storia della Fisica e dell' Astronomia*, 229-249.

Dalzell, D.P. (1930). 'Heaviside's operational method', *Proceedings of the Physical Society (1926-1948)*, 42(2): 75-81.

Danos, M. (1982). 'Bohm-Aharonov effect: The quantum mechanics of the electrical transformer', *American Journal of Physics*, **50**(1): 64-66.

Darrieus, G. (1970). 'Puissance réactive et action', *Revue Générale de l'Électricité*, **79**(9): 701-707.

Darrigol, O. (1996). 'The electrodynamic origins of relativity theory', *Historical Studies in the Physical and Biological Sciences*, **26**(2): 241-312.

Darrigol, O. (2000a). 'Poincaré, Einstein, et l'inertie de l'énergie', *Comptes rendus de l'Académie des Sciences*, Series, 4, 1: 143-153.

Darrigol, O. (2000b). *Electrodynamics from Ampère to Einstein*. Oxford: Oxford University Press.

Davis, M. (1994). 'A unified theory of lumped circuits and differential systems based on Heaviside operators and causality', *IEEE Transactions on Circuits and Systems – I: Fundamental Theory and Applications*, **41**(11): 712-727.

Davis, B.S. and L. Kaplan (2011). 'Poynting vector in a circular circuit', *American Journal of Physics*, **79**(11): 1155-1162.

Dawson, J.W. (1997). *Logical Dilemmas. The Life and Work of Kurt Gödel*. Wellesley: A.K. Peters, Ltd.

Deakin, M.A.B. (1992). 'The ascendancy of the Laplace Transform and how it came about', *Archive for History of Exact Sciences*, **44**(3): 265-286.

De Broglie, L. [1923] (2006). 'A tentative theory of light quanta', *Philosophical Magazine Letters*, **86**(7): 411-423. https://doi.org/10.1080/09500830600914721.

De Broglie, L–V. [1925] (2004). *On the theory of Quanta*. English translation by A.F. Kracklauer of '*Recherches sur la théorie des quanta*', *Annales de Physique*, 10th series, 3 (Jan.-Feb. 1925).

Deckert, D.-A. and V. Hartenstein (2016). 'On the initial value formulation of classical electrodynamics', *Journal of Physics A: Mathematical and Theoretical*, **49**(44): 445202.

de Leon, F. and J. Cohen. (2008). Discussion of 'Instantaneous reactive power p-q theory and power properties of three-phase systems', *IEEE Transactions on Power Delivery*, **23**(3): 1693-1694.

de Leon, F. and J. Cohen (2010). 'AC power theory from Poynting theorem: Accurate identification of instantaneous power components in nonlinear-switched circuits', *IEEE Transactions on Power Delivery*, **25**(4): 2104-2112.

Della Torre, F., A. P. Morando and G. Sapienza (2010). 'Cisoidal algebra: Some meanings for a unified energetic balance', *Proceedings of 14th International Conference on Harmonics and Quality of Power - ICHQP 2010*, Bergamo: 1-4. https://doi.org/10.1109/ICHQP.2010.5625381.

Demir, S., M. Tanish and N. Candemir (2010). 'Hyperbolic quaternion formulation of electromagnetism', *Advances in Applied Clifford Algebras*, **20**(3): 547-563.

Denzel, P. (1966). *Grundlagen der Übertragung elektrischer Energie*. Berlin: Springer.

Depenbrock, M. (2001). 'Variation power, variation currents: Physical background and compensation rules', *European Transactions on Electrical Power*, 11(5): 309-316.

Depenbrock, M. and V. Staudt (1998). 'Hyper Space Vectors: A New Four-Quantity Extension of Space-Vector Theory', *European Transactions on Electrical Power*, 8(4): 241-248.

Deshpande, M.V. (1970). *Electrical Power System Design*. New Delhi: Tata McGraw-Hill Publishing Co. Limited.

Despotovic, S. (1962). *Osnovi analize elektro energetskih sistema*. Belgrade: Zajednica Jugoslovenske Elektroprivrede.

Dhar, R.N. (1983). *Computer-aided Power System Operation & Analysis*. New Delhi: Tata McGraw-Hill Publishing Co. Limited.

Dirac, P.A.M. (1963). 'The evolution of the physicist's picture of nature', *Scientific American*, 208(5): 45-53.

Doetsch, G. (1943). *Theorie und Anwendung der Laplace-Transformation*. New York: Dover Publications.

Doncel, M.G. and J.A. de Lorenzo (1996). 'The electrotonic state, a metaphysical device for Maxwell too?' *European Journal of Physics*, 17(1): 6-10.

Doran, C. and A. Lasenby (2003). *Geometric Algebra for Physicists*. Cambridge University Press.

Dorier, J-L. (1995). 'A general outline of the genesis of vector space theory', *Historia Mathematica*, 22(3): 227-261.

Dorn, R.T. (2009). 'A two-body photon model', *Proceedings of the Society of Photo-Optical Instrumentation Engineers (SPIE) 7421, The Nature of Light: What are Photons?* III, 742119.

Dorn, R.T. (2015). 'Electron-positron annihilation and absorption models', *Proceedings of the Society of Photo-Optical Instrumentation Engineers (SPIE) 9570, The Nature of Light: What are Photons?* VI, 95700H.

Dorst, L., D. Fontijne and S. Mann (2007). *Geometric Algebra for Computer Science. An Object-oriented Approach to Geometry*. Amsterdam: Elsevier, Morgan Kaufmann Publ.

Dressel, J., K.Y. Bliokh and F. Nori (2015). 'Spacetime algebra as a powerful tool for electromagnetism', *Physics Reports*, 589: 1-71.

Drude, P. (1900a). 'Zur Elektronen theorie der Metalle', *Annalen der Physik*, 306(3): 566-613.

Drude, P. (1900b). 'Zur Elektronen theorie der Metalle: II Teil. Galvanomagnetische und thermomagnetische Effecte', *Annalen der Physik*, 308(11): 369-402.

Duhem, M.P. (1902). *Les théories électriques de J. Clerk Maxwell. Étude Historique et Critique*. Paris: Librairie Scientifique A. Hermann.

Dunmore, C. (1991). 'Meta-level revolutions in mathematics'. In: *Revolutions in Mathematics*, (Ed.). D. Gillies, 209-235. Oxford: Clarendon Press.

Dyson, F.J. (1972). 'Missed opportunities', *Bulletin of the American Mathematical Society*, **78**(5): 635-652.

Eckmann, B. (1943). 'Stetige Lösungen linearer Gleichungssysteme' (Continuous solutions of linear systems of equations), *Commentarii Mathematici Helvetici*, 15: 318-339.

Edelmann, H. (1981). 'Wirkleistung, Blindleistung, Scheinleistung bei periodischen Strömen und Spannungen in funktionsanalytischer Sicht', *Siemens Forschungs und Entwicklungs berichte*, **10**(1): 16-24.

Edminister, J.A. (1983). *Theory and Problems of Electric Circuits*. New York: McGraw-Hill.

Electrical Engineering Staff, M.I.T. (1948). *Electric Circuits. A First Course in Circuit Analysis for Electrical Engineers*.

Einstein, A. (1905). 'Zur Elektrodynamik bewegter Körper', *Annalen der Physik*, **322**(10): 891-921.

Einstein, A., B. Podolsky and N. Rosen (1935). 'Can quantum-mechanical description of physical reality be considered complete?', *Physical Review*, **47**(10): 777-780.

Eisenbud, D. (2005). *The Geometry of Syzygies: A Second Course in Commutative Algebra and Algebraic Geometry*. New York: Springer.

Elgerd, O.I. (1982). *Electric Energy Systems Theory. An Introduction*. New York: McGraw-Hill, Inc.

Emanuel, A.E. (1974).'Suggested definition of reactive power in non-sinusoidal systems', *Proceedings of the Institution of Electrical Engineers*, **121**(7): 705-706.

Emanuel, A.E. (1977). 'Energetical factors in power systems with nonlinear loads', *Archiv für Elektrotechnik*, 59:183-189.

Emanuel, A.E. (1990a). 'Power components in a system with sinusoidal and non-sinusoidal voltages and/or currents', *IEE Proceedings B (Electric Power Applications)*, **137**(3): 194-196.

Emanuel, A.E. (1990b). 'Powers in nonsinusoidal situations. A review of definitions and physical meaning', *IEEE Transactions on Power Delivery*, **5**(3): 1377-1383.

Emanuel, A.E. (1993). 'Apparent and reactive powers in three-phase systems: In search of a physical meaning and a better resolution', *European Transactions on Electrical Power*, **3**(1): 7-14.

Emanuel, A.E. (1996). 'The oscillatory nature of the power in single- and polyphase circuits', *European Transactions on Electrical Power* **6**(5): 315-320.

Emanuel, A.E. (2004). 'Summary of IEEE Standard, 1459: Definitions for the measurement of electric power quantities under sinusoidal, non-sinusoidal, balanced, or unbalanced conditions', *IEEE Transactions on Industry Applications*, **40**(3): 869-876.

Emanuel, A.E. (2005). 'Poynting vector and the physical meaning of non-active powers', *IEEE Transactions on Instrumentation and Measurement*, **54**(4): 1457-1462.

Emanuel, A.E. (2010). *Power Definitions and the Physical Mechanism of Power Flow*, IEEE Press. Chichester, U.K.: Wiley Ltd.

Emanuel, A.E. and J.A. Orr (2012). 'Fryze's power definition: Some limitations', *2012 IEEE 15th International Conference on Harmonics and Quality of Power*, 518-522. https://doi.org/10.1109/ICHQP.2012.6381170.

Emde, F. (1902). *Die Arbeitsweise der Wechselstrom-maschinen*. Berlin: Springer.

Emde, F. (1923). 'Polare und axiale Vektoren in der Physik', *Zeitschrift für Physik A. Hadrons and Nuclei*, **12**(1): 258-264.

Engelhardt, W.W. (2016). 'Ohm's law and Maxwell's equations', *Annales de la Fondation Louis de Broglie*, **41**(1): 39-53.

Enk, van S.J. (2013). 'The covariant description of electric and magnetic field lines and null fields: Application to Hopf-Rañada solutions' 13 p, arXiv: 1302.2683v2 [physics. optics] 25 Mar. 2013.

Enslin, J.H.R. (1988). *Determination and Dynamic Compensation of Fictitious Power in Electric Power Systems*, thesis for the degree of Doctor in Engineering, Rand Afrikaans University, Pretoria, South Africa, June 1988.

Enslin, J.H.R. and J.D. van Wyk (1988). 'Digital signal processing in electrical power systems: Calculation of power under non-sinusoidal voltage and current conditions', *The Transactions of the South African Institute of Electrical Engineering*, **79**(1): 17-24.

Enslin, J.H.R. (1990). 'Orthogonal representation of power with the aid of quaternions', *Suid-Afrikaanse Tydskrif vir Natuurwetenskap en Tegnologie*, **9**(1): 11-14.

Epstein, P.S. (1914). 'Die ponderomotorischen Drehwirkungen einer Lichtwelle und die Impulssätze der Elektronentheorie', *Annalen der Physik*, **349**(12): 593-604.

Erdélyi, A. (1959). *Operational Calculus and Generalized Functions*. Pasadena: California Institute of Technology.

Eremia, M. (2006). *Electric Power Systems, Electric Networks*. Bucharest: *Editura Academiei Române*.

Erlicki, M.S. and A. Emanuel-Eigeles (1968). 'New aspects of power factor improvement. Part I – Theoretical Basis', *IEEE Transactions on Industry and General Applications*, IGA, **4**(4): 441-446.

Estwick, C.F. (1953). 'Real power and imaginary power in a-c circuits', *Transactions of the American Institute of Electrical Engineers. Part III: Power Apparatus and Systems*, **72**(1): 27-35.

Evans, M.W. (2001). 'The link between the Sachs and O (3) theories of electrodynamics'. In: *Modern Nonlinear Optics, Part 2*, 2nd ed., (Ed.). M.W. Evans, 469-494, *Advances in Chemical Physics*, 119. Wiley.

Everett, H., III. (1973). 'The theory of the universal wave theory'. In: *The Many Worlds Interpretation of Quantum Mechanics*, (Ed.). B. deWitt and N. Graham, 1-140. Princeton: Princeton University Press.

Exposito, A.G. (2002). *Analisis y operacion de sistemas de energia eléctrica*. Madrid: McGraw Hill.

Familton, J.C. (2015). *Quaternions: A History of Complex Noncommutative Rotation Groups in Theoretical Physics*, PhD thesis, Columbia University, UMI Number 3702435.

Fano, R.M., L.J. Chu and R.B. Adler (1968). *Electromagnetic Fields, Energy, and Forces*. Cambridge: The MIT Press.

Faria, J.A.B. (2008). *Electromagnetic Foundations of Electrical Engineering*. Wiley.

Faria, J.A.B. (2009). 'The role of Poynting's vector in polyphase power calculations', *European Transactions on Electrical Power*, **19**(5): 683-688.

Faria, J.A.B. (2012). 'Poynting vector flow analysis for contactless energy transfer in magnetic systems', *IEEE Transactions on Power Electronics*, **27**(10): 4292-4300.

Faria, J.A.B. (2013). 'A physical model of the ideal transformer based on magnetic transmission line theory', *Journal of Electromagnetic Waves and Applications*, **27**(3): 365-373.

Farouki, R.T. (2008). *Pythagorean-Hodograph Curves: Algebra and Geometry inseparable*. Berlin: Springer.

Favaro, A., F.W. Hehl and J. Lux (2016). 'On the metamorphoses of Maxwell's equations during the last 150 years: Spotlights on the history of classical electrodynamics', *2016 URSI International Symposium on Electromagnetic Theory (EMTS)* (Espoo, Finland): 306-307. https://doi. org/10.1109/URSI-EMTS.2016.7571381.

Fearnley-Sander, D. (1979). 'Hermann Grassmann and the creation of linear algebra', *The American Mathematical Monthly*, **86**(10): 809-817.

Felsberg, M. and G. Sommer (2001). 'The monogenic signal', *IEEE Transactions on Signal Processing*, **49**(12): 3136-3144.

Felsen, L.B., M. Mongiardo and P. Russer (2009). *Electromagnetic Field Computation by Network Methods*. Berlin: Springer.

Ferguson, O.J. (1903). 'Quaternions in electrical calculations', *Physical Review* (Series 1). **17**(5): 378-381.

Ferraris, G. [1887] (1901). 'Sulle differenze di fase delle correnti, sul ritardo dell'induzione e sulla dissipazione di energia nei transformatori', *Memoria presentata alla Real Accademia delle Scienze di Torino il*, 4 Dec. 1887. In: *Opere di Galileo Ferraris*, vol. 1, 261-323, 1902. *Ulrico Hoepli Editore-Libraio della Real Casa*: 1902.

Ferraris, G. (1888a). 'Sulle differenze di fase delle correnti, sul ritardo dell'induzione e sulla dissipazione di energia nei trasformatori', *Il Nuovo Cimento*, **23**(1): 193-211.

Ferraris, G. (1888b). 'Rotazioni elettrodinamiche prodotte per mezzo di correnti alternate; Nota del Prof. Galileo Ferraris', *Il Nuovo Cimento*, **24**(1): 242-256.

Ferraris, G. (1894). 'Un metodo per la trattazione dei vettori rotanti od alternative ed una applicazione di esso ai motori elettrici a correnti alternate', *Il Nuovo Cimento*, **35**(1): 99-126.

Ferreira, J.A. (1988). 'Application of the Poynting vector for power conditioning and conversion', *IEEE Transactions on Education*, **31**(4): 257-264.

Ferrero, A. (1996). 'Mathematical foundations of the instantaneous power concepts: An algebraic approach', *European Transactions on Electrical Power*, **6**(5): 305-309.

Ferrero, A. (2008). 'Measuring electric power quality: Problems and perspectives', *Measurement*, **41**(2): 121-129.

Ferrero, A., L. Giuliani and J.L. Willems (2000). 'A new space-vector transformation for four-conductor systems', *European Transactions on Electrical Power*, **10**(3): 139-145.

Ferrero, A., S. Leva and A.P. Morando (2001). 'An approach to the non-active power concept in terms of the Poynting-Park vector', *European Transactions on Electrical Power*, **11**(5): 291-299.

Ferrero, A., S. Leva and A.P. Morando (2004). 'A systematic, mathematically and physically sound approach to the energy in three-wire, three-phase systems', *L'EnergiaElettrica*. 81: 51-56.

Fetea, R. and A. Petroianu (2000a). 'Reactive power: A strange concept', *Second European Conference on Physics Teaching in Engineering Education (PTEE 2000)*, 14-17 June, Budapest University of Technology and Economics, Hungary.

Fetea, R. and A. Petroianu (2000b). 'Can the reactive power be used?' PowerCon, 2000 (2000 International Conference on Power System Technology, 4-7 December 2000, The University of Western Australia, Perth, Australia), *IEEE Catalog Number: 00EX409*, 1251-1256.

Feynman, R. (1961). *Quantum Electrodynamics*. Reading, MA: Addison-Wesley.

Feynman, R. (1966). 'The Development of the Space-Time View of Quantum Electrodynamics', *Science*, **153**(3737): 699-708.

Feynman, R., R.B. Leighton and M. Sands (1964). *The Feynman Lectures on Physics*, vol. 2: *Mainly Electromagnetics and Matter*. Reading, MA: Addison-Wesley.

Fich, S. and J.L. Potter (1958). *Theory of A-C Circuits*. Englewood Cliffs, NJ: Prentice-Hall.

Filipski, P. (1980). 'A new approach to reactive current and reactive power measurement in non-sinusoidal systems', *IEEE Transactions on Instrumentation and Measurement*, **29**(4): 423-426.

Filipski, P. (1993). 'Apparent power-a misleading quantity in the non-sinusoidal power theory: Are all non-sinusoidal power theories doomed to fail?' *European Transactions on Electrical Power*, **3**(1): 21-26.

Filipski, P., Y. Baghouz and M.D. Cox (1994). 'Discussion of power definitions contained in the IEEE dictionary', *IEEE Transactions on Power Delivery*, **9**(3): 1237-1243.

Filipski, P. and P.W. Labaj (1992). 'Evaluation of reactive power meters in the presence of high harmonic distortion', *IEEE Transactions on Power Delivery*, **7**(4): 1793-1799.

Fink, D.G. and H.W. Beaty (1987). *Standard Handbook for Electrical Engineers*,12th ed. New York: McGraw-Hill.

Fjelstad, P. (1986). 'Extending relativity via perplex numbers', *American Journal of Physics*, **54**(5): 416-422.

Fjelstad, P. and S.S. Gal (2001). 'Two-dimensional geometries, topologies, trigonometries and physics generated by complex-type numbers', *Advances in Applied Clifford Algebras*, **11**(1): 81-107.

Flegg, H.G. (1971). 'A survey of the development of operational calculus', *International Journal of Mathematical Education in Science and Technology*, **2**(4): 329-335.

Flegg, H.G. (1974). 'Mikusinski's Operational Calculus', *International Journal of Mathematical Education in Science and Technology*, **5**(2): 131-137.

Fleming, J.A. (1887). 'Notes on alternating currents', *The Electrician* (November 18) : 28-30.

Fleming, J.A. (1911). 'Electricity', *The Encyclopaedia Britannica*, 11th ed., vol. 9: 192.

Flood, R., M. McCartney and A. Whitaker (Eds.). (2014). *James Clerk Maxwell: Perspectives on His Life and Work*. Oxford University Press.

Fölsing, A. (1997). *Heinrich Hertz: Eine Biographie*. Hamburg: Hoffmann und Campe.

Föppl, A. (1894). *Einführung in die Maxwell'sche Theorie der Elektricität: mit einen einleitenden Abschnitte über das Rechnen mit Vektorgrössen in der Physik*. Leipzig: Druck und Verlag von B.G. Teubner.

Fortescue, C.L. (1933). 'Power, reactive volt-amperes, power factor', *Transactions of the American Institute of Electrical Engineers*, **52**(3): 758-762.

Fradkin, E. (2013). *Field Theories of Condensed Matter Physics*, 2nd ed., Cambridge: Cambridge University Press.

Fradkin, E.S. and M.Ya. Palchik (1996). 'Conformal Quantum Field Theory in D-Dimensions', *Mathematics and Its Applications* 376. Kluwer Academic Publishers.

Francis, M.R. and A. Kososwsky (2005). 'The construction of spinors in geometric algebra', *Annals of Physics*, 317: 383-409.

Franklin, W.S. (1901). 'Poynting's theorem and the distribution of electric field inside and outside of a conductor carrying electric current', *Physical Review* (Series 1), **13**(3): 165-181.

Franklin, W.S. (1903). 'A discussion of some points in alternating current theory', *Transactions of the American Institute of Electrical Engineers*, 21: 589-601.

Franklin, W.S. (1912). 'Poynting's theorem and the equations of electromagnetic action', *Journal of the Franklin Institute*, **173**(1): 49-54.

Freundlich, Y. (1978). 'In defence of Copenhagenism', *Studies in History and Philosophy of Science Part A*, **9**(3): 151-179.

Frisch, M. (2004). 'Inconsistency in classical electrodynamics', *Philosophy of Science*, **71**(4): 525-549.

Frisch, M. (2005). *Inconsistency, Asymmetry, and Non-Locality. A Philosophical Investigation of Classical Electrodynamics.* Oxford University Press.

Frisch, M. (2008). 'Conceptual problems in classical electrodynamics', *Philosophy of Science*, **75**(1): 93-105.

Frisch, M. (2009). 'Philosophical issues in electromagnetism', *Philosophy Compass*, **4**(1): 255-270.

Frisch, M. (2014). 'Models and scientific representations or: who is afraid of inconsistency?' *Synthese*, **191**(13): 3027-3040.

Fry, T.C. (1929). 'Differential equations as a foundation for electrical circuit theory', *The American Mathematical Monthly*, **36**(10): 499-504.

Fryze, S. (1932). 'Wirk-, Blind- und Scheinleistung in elektrischen Stromkreisen mit nicht-sinusförmigem Verlauf von Strom und Spannung', *Elektrotechnische Zeitschrift* (ETZ), **53**(23): 596-599; (25): 625-627 (26): 700-702.

Furga, S.G. and L. Pinola (1994). 'The mean generalized content: A conservative quantity in periodic-forced non-linear networks', *European Transactions on Electrical Power Engineering*, **4**(3): 205-212.

Gabor, D. (1946). 'Theory of communication. Part 1: The analysis of information; Part 2: The analysis of hearing; Part 3: Frequency compression and expansion', *Journal of the Institution of Electrical Engineers - Part III: Radio and Communication Engineering*, **93**(26): 429-457.

Gaiceanu, M. (2005). 'Active power compensator of the current harmonics based on the instantaneous power theory', *The Annals of "Dunărea de Jos. University of Galaţi*, Fascicle III: 23-28.

Galili, I. and D. Kaplan (1997). 'Changing approach to teaching electromagnetism in a conceptually oriented introductory physics course', *American Journal of Physics*, **65**(7): 657-667.

Galili, I. and E. Goihbarg (2005). 'Energy transfer in electrical circuits: A qualitative account', *American Journal of Physics*, **73**(2): 141-144.

Galili, I., D. Kaplan and Y. Lehavi (2006). 'Teaching Faraday's law of electromagnetic induction in an introductory physics course', *American Journal of Physics*, **74**(4): 336-343.

Gamba, A. (1967). 'Peculiarities of the eight-dimensional space', *Journal of Mathematical Physics*, **8**(4): 775-781.

Garrigues-Baixauli, J. (2019). 'Discrete Model of Electron', *Applied Physics Research*, **11**(6): 36-55.

Garrity, T.A. (2015). *Electricity and Magnetism for Mathematicians. A Guided Path from Maxwell's Equations to Yang-Mills*. Cambridge University Press.

Genkin, M. (1926). 'Expression de la puissance', *Revue Générale de l'Électricité*, **20**(26): 963-969.

Gerbracht, E.H.A. [1998] (2007). 'An extension to an algebraic method for linear time-invariant system and network theory: The full AC-Calculus',arXiv:0709.2935v1 [math.CA] 19 Sep. 2007.

Ghassemi, F. (2000a). 'New concept in AC power theory', *IEE Proceedings C – Generation, Transmission and Distribution*, **147**(6): 417-424.

Ghassemi, F. (2000b). 'New apparent power and power factor with non-sinusoidal waveforms', *2000 IEEE Power Engineering Society Winter Meeting (Singapore), Conference Proceedings* 4, 2852-2857. https://doi.org/10.1109/PESW.2000.847337.

Ghassemi, F. (2004). 'Electrical Power Measurement' United States Patent No.: US 6,828,771 B1, Dec. 7, 2004. 28 p.

Gibbs, J.W. (1881-1884). *Elements of Vector Analysis*. New Haven: Tuttle, Morehouse & Taylor. https://doi.org/10.5479/sil.325602.39088000942730.

Gibbs, J.W. (1886). 'On multiple algebra', *Proceedings of the American Association for the Advancement of Science*, Salem, MA: 1-32.

Gibbs, J.W. (1891). 'Quaternions and the *Ausdehnungslehre*', *Nature*, **44**(1126: 79-82.

Gibbs, J.W. (1907). *Vector Analysis. A Text-book for the Students of Mathematics and Physics. Founded upon the lectures of J. Willard Gibbs*, by E.B. Wilson. New York: Charles Scribner's Sons.

Gibbs, W.J. (1952). *Tensors in Electrical Machine Theory*. London: Chapman & Hall Ltd.

Giblin, S.P. (2011). 'Electron pumps and redefinition of the SI unit Ampere', XXXthURSI [Union Internationale de Radio-Scientifique] *General Assembly and Scientific Symposium*, Istanbul, 13-20 August. https://doi.org/10.1109/URSIGASS.2011.6123711

Gillies, D. (Ed.). (1991). *Revolutions in Mathematics*. Oxford: Clarendon Press.

Gingras, Y. (1980)]. Comment on 'What the electromagnetic vector potential describes', *American Journal of Physics*, **48**(1): 84.

Girard, P.R. (1984). 'The quaternion group and modern physics', *European Journal of Physics*, **5**(1): 25-32.

Girard, P.R. (2007). *Quaternions, Clifford Algebras and Relativistic Physics.* Basel: Birkhäuser.

Glaeske, H.-J., A.P. Prudnikov and K.A. Skornik (2006). *Operational Calculus and Related Topics.* Chapman & Hall/CRC.

Glashow, S.L. (1980). 'Toward a Unified Theory: Threads in a Tapestry', *Science*, **210**(4476): 1319-1323.

Glover, J.D. and M.S. Sarma (2002). *Power System Analysis and Design*, 3rd ed., Pacific Grove, CA: Brooks/Cole Thomson Learning.

Gluckman, A.G. (1999). 'On W.E. Weber's *Grundprincip der Elektrodynamik* and related earlier and contemporary studies', *Physics Essays*, **12**(4): 682-698.

Gluskin, E. (1997). 'Nonlinear systems: Between a law and a definition', *Reports on Progress in Physics*, 60: 1063-1112.

Gluskin, E. (2010). 'The electromagnetic 'memory' of a dc-conducting resistor: A relativity argument and the electrical circuits',arXiv:1005.0997v6 [math-ph] 20 June: 1-10.

Goedecke, G.H. (2001). 'On energy absorption in classical Electromagnetism', *American Journal of Physics*, **69**(2): 226-228.

Goenner, H. (2017). Some remarks on 'A contribution to Electrodynamics' by Bernhard Riemann'. In: *From Riemann to Differential Geometry and Relativity* (Ed.). L. Ji, A. Papadopoulos and S. Yamada, 111-123. Springer International Publishing AG.

Gogberashvili, M. (2005). 'Octonionic Geometry', *Advances in Applied Clifford Algebras*, **15**(1): 55-66.

Gönen, T. (1988). *Modern Power System Analysis.* New York: Wiley.

González Calvet, R.G. (2007). *Treatise of Plane Geometry through Geometric Algebra*, TIMSAC series on applied mathematics 1. *Cerdanyola del Vallès*, Spain: *Taller d'Integració Matemàtica en els Sabers Compartits.*

González Calvet, R.G. (2010). 'Applications of geometric algebra and the geometric product to solve geometric problems', *Applied Geometric Algebras in Computer Science and Engineering (AGACSE) Conference* held in Amsterdam on 14-16 June, 2010. 14 p.

González Calvet, R. (2013). 'New foundations for geometric algebra', *Electronic Journal: Clifford Analysis, Clifford Algebras and Their Applications*, **2**(3): 193-211.

Gooding, D. (1978). 'Conceptual and experimental bases of Faraday's denial of electrostatic action at a distance', *Studies in the History and Philosophy of Science*, **9**(2): 117-149.

Gori, F., S. Vicalvi, M. Santarsiero, F. Frezza, G. Schettini, S. Ambrosini and R. Borghi (1997). 'An elementary approach to spinors', *European Journal of Physics*, **18**(4): 256-262.

Gough, W. (1982). 'Poynting in the wrong direction', *European Journal of Physics*, **3**(2): 83-87.

Grabinski, H. and F. Wiznerowicz (2010). 'Energy transfer on three-phase high-voltage lines; the strange behavior of the Poynting vector', *Electrical Engineering*, **92**(6): 203-214.

Grainger, J.J. and W.D. Stevenson (1994). *Power System Analysis*. New York: McGraw-Hill.

Graneau, N., Phipps Jr, T. & Roscoe, D. (2001). 'An experimental confirmation of longitudinal electrodynamic forces', *European Physical Journal D – Atomic, Molecular, Optical and Plasma Physics*, 15: 87-97. https://doi.org/10.1007/s100530170186.

Graneau, P. (1984). 'Ampere tension in electric conductors', *IEEE Transactions on Magnetics*, **20**(2): 444-455.

Graneau, P. (1991). 'Non-local action in the induction motor', *Foundations of Physics Letters*, **4**(5): 499-506.

Gray, J. (1979). *Ideas of Space: Euclidean, Non-Euclidean, and Relativistic*. Oxford: Clarendon Press.

Gray, J. (1980). 'Olinde Rodrigues' paper of 1840 on transformation groups', *Archive for History of Exact Sciences*, **21**(4): 375-385.

Griffiths, D.J. (2012). 'Resource Letter EM-1: Electromagnetic Momentum', *American Journal of Physics*, **80** (1): 7-18.

Griffiths, D.J. (2013). *Introduction to Electrodynamics*. Boston: Pearson.

Griffiths, D.J., T.C. Proctor and D.F. Schroeter (2010). 'Abraham-Lorentz versus Landau-Lifshitz', *American Journal of Physics*, **78**(4): 391-402.

Grigsby, L.L. (1998). *The Electric Power-Engineering Handbook*. New York: IEEE Press.

Grimes, C.A. and D.M. Grimes (1997a). 'Complex power in circuits with multiple reactive elements', *Electric Machines and Power Systems*, **25**(9): 955-966.

Grimes, C.A. and D.M. Grimes (1997b). "The Poynting theorems and the potential for electrically small antennas', *1997 IEEE Aerospace Conference*, 3: 161-176.

Grimes, D.M. and C.A. Grimes (1995). 'The complex Poynting Theorem. Reactive power, radiative Q, and limitations on electricity small antennas', *Proceedings of International Symposium on Electromagnetic Compatibility* (14-18 August): 97-101.

Grimes, D.M. and C.A. Grimes (1997). 'Power in modal radiation fields: Limitations of the complex Poynting theorem and the potential for electrically small antennas', *Journal of Electromagnetic Waves and Applications*, **11**(12): 1721-1747.

Grivoyannis, B.H. (2014). *Of pirates and Propellers: Complex Numbers in Plane Geometry*, Master's thesis. Kean University, Union, New Jersey.

Gröchenig, Karlheinz (2001). *Foundations of Time-Frequency Analysis*. Boston: Birkhäuser.

Gross, C.A. (2006). 'On VA's, VAR's, and other traditions in electric power engineering', *Proceedings of the 38th North American Power Symposium (NAPS)*, 2006: 383-387.

Grünholz, H. (1928). *Theorie der Wechselstromübertragung (Fernleitung und Umspannung)*. Berlin: Verlag von Julius Springer.

Guarnieri, M. (2011). 'A straightforward deduction of the electric circuit power', *COMPEL, The International Journal for Computation and Mathematics in Electrical and Electronic Engineering*, **30**(4): 1271-1282.

Guilbert, C.-F. (1900). 'Représentation des fonctions périodiques complexes', *L'Éclairage Électrique*, **22**(11): 405-414.

Guile, A.E. and W. Paterson (1977). *Electrical Power Systems*, vol. 1. Oxford: Pergamon Press.

Guillemin, E.A. (1953). *Introductory Circuit Theory*. Oxford: Pergamon Press.

Gull, S., A. Lasenby and C, Doran (1993). 'Imaginary numbers are not real – The geometric algebra of space-time', *Foundations of Physics*, **23**(9): 1175-1201.

Gullberg, J. (1997). *Mathematics: From the Birth of Numbers*. New York: W.W. Norton & Co.

Gürlebeck, K., K. Habetha and W. Sprößig (2008). *Holomorphic Functions in the Plane and N-dimensional Space*. Birkhäuser, Basel.

Hadfield, H., D. Hildebrand and A. Arsenovic (2019). 'GAJIT: Symbolic optimisation and JIT compilation of geometric algebra in Python with GAALOP and Numba'. In: *Advances in Computer Graphics, Proceedings of the 36th Computer Graphics International Conference (CGI 2019)* held in Calgary, AB, Canada, June 17–20 2019, (Ed.). M. Gavrilova, J. Chang, N. Magnenat-Thalmann, E. Hitzer and H. Ishikawa, 499-510. Springer Lecture Notes in Computer Science.

Hadamard, J. (1945). *The Psychology of Invention in the Mathematical Field*. New York: Dover Publications.

Haddad, D., F. Seifert, L.S. Chao, S. Li., D.B. Newell, *et al.* (2016). 'Bridging classical and quantum mechanics', *Metrologia*, 53: A83-A85.

Hahn, S.L. (1996). *Hilbert Transforms in Signal Processing*. Norwood, MA: Artech House.

Hahn, S.L. (2007). 'The history of applications of analytic signals in electrical and radio engineering', EUROCON 2007: *The International Conference on Computer as a Tool*, Warsaw, 2627-2631.

Hahn, S.L. (2010). 'Hilbert Transforms'. In: *Transforms and Applications Handbook*, (Ed.). A.D. Poularikas, Chapter 7 (7-2-7-99). Boca Raton, FL: CRC Press.

Hahn, S.L. and K.M. Snopek (2017). *Complex and Hypercomplex Analytic Signals: Theory and Applications*. Boston: Artech House.

Halevi, P. (1980). 'Plane electromagnetic waves in material media: Are they transverse waves?', *American Journal of Physics*, **48**(10): 861-867.

Halevi, P. and A. Mendoza-Hernandez (1981). 'Temporal and spatial behavior of the Poynting vector in dissipative media: Refraction from vacuum into medium', *Journal of the Optical Society of America*, **71**(10): 1238-1242.

Hamar, R., L. Sroubova and P. Kropic (2014). 'Electromagnetic field along the power overhead line at point where the line route changes direction', *COMPEL, The International Journal for Computation and Mathematics in Electrical and Electronic Engineering*, **33**(6): 1950-1964.

Hammond, P. (1958). *Electromagnetic Energy Transfer*, The Institution of Electrical Engineers, Monograph No. 286, Feb.: 352-358.

Hammond, P. (1981). *Energy Methods in Electromagnetism*. Oxford: Clarendon Press.

Hammond, P. (1999). 'The role of potentials in electromagnetism', *COMPEL, The International Journal for Computation and Mathematics in Electrical and Electronic Engineering*, **18**(2): 103-119.

Handschin, E. (1987). *Elektrische Energie-Übertragungssysteme*. Heidelberg: Dr Alfred Hüthig Verlag.

Hankel, H. (1867). *Vorlesungen über die Complexen Zahlen und ihre Funktionen*. Leipzig: Leopold Voss.

Happoldt, H. and D. Oeding (1978). *Elektrische Kraftwerke und Netze*. Berlin: Springer.

Harbola, M.K. (2010). 'Energy flow from a battery to other circuit elements: Role of the surface charges', *American Journal of Physics*, **78**(11): 1203-1206.

Harkin, A.A. and J.B. Harkin (2004). 'Geometry of generalized complex numbers', *Mathematics Magazine*, **77**(2): 118-129.

Harman, P.M. (1982). *Energy, Force, and Matter: The Conceptual Development of Nineteenth-century Physics*. Cambridge: Cambridge University Press.

Harman, P.M. (1993). 'Maxwell and Faraday', *European Journal of Physics*, **14**(4): 148-154.

Harmuth, H.F. (1986a). 'Correction for Maxwell's equations for signals I', *IEEE Transactions on Electromagnetic Compatibility* EMC, **28**(4): 250-256.

Harmuth, H.F. (1986b). 'Correction for Maxwell's equations for signals II', *IEEE Transactions on Electromagnetic Compatibility* EMC, **28**(4): 259-265.

Harmuth, H.F. (1986c). 'Propagation velocity of electromagnetic signals', *IEEE Transactions on Electromagnetic Compatibility* EMC, **28**(4): 267-271.

Harmuth, H.F. (1986d). *Propagation of Non-sinusoidal Electromagnetic Waves*, Orlando, FL: Academic Press. 245 p.

Harmuth, H.F. (1989). 'Reply on Comments on *Maxwell's equations* by J.E. Gray and S.P. Bowen', *IEEE Transactions on Electromagnetic Compatibility*, **31**(2): 200.

Harmuth, H.F. (1991). 'Hillion's proof of the nonexistence of certain solutions of Maxwell's equations', *IEEE Transactions on Electromagnetic Compatibility*, **33**(4): 371-372.

Harmuth. H.F. (2001). *Modified Maxwell Equations in Quantum Electrodynamics*. World Scientific Publishing Company.

Harmuth, H.F. and M.G.M. Hussain (1994). 'Signal Solutions of Maxwell's Equations for Charge Carriers with Non-Negligible Mass', *IEEE Transactions on Electromagnetic Compatibility*, **36**(4): 411-412.

Harpaz, A. (2002). 'The nature of fields', *European Journal of Physics*, **23**(3): 263-268.

Harrer, B.W. (2017). 'On the origin of energy: Metaphors and manifestations as resources for conceptualizing and measuring the invisible, imponderable', *American Journal of Physics*, **85**(6): 454-460.

Harrington, R.F. (2001). *Time-harmonic Electromagnetic Fields*. Piscataway, NJ: IEEE.

Harrison, D. (1981). 'Discovering induction', *American Journal of Physics*, **49**(6): 603.

Hartenstein, V. (2018). *On the Maxwell-Lorentz dynamics of Point Charges*, dissertation, Ludwig Maximilians Universität München, München.

Hartenstein, V. and M. Hubert (2018). 'When fields are not degrees of freedom'. PhilSci Archive http://philsci-archive.pitt.edu/id/eprint/14911.

Hartman, M.T. (2007). 'The application of Fortescue's transformation to describe power states in multi-phase circuits with non-sinusoidal voltage and currents', *9th International Conference Electrical Power Quality and Utilisation*, Barcelona, 9-11 October.

Hartman, M.T. and M. Hashad (2008a). 'The correlation functions of power as a new proposition to describe power states in circuits with periodical voltage and current waveforms', *International School on Non-sinusoidal Currents and Compensation*, Lagow, Poland: 1-4.

Hartman, M.T. and M. Hashad (2008b). 'A few remarks on the analysis of energy transfer through any periodic current and voltage waveforms', *International School on Non-sinusoidal Currents and Compensation*, Lagow, Poland: 1-5.

Haus, H.A. (1969). 'Momentum, energy and power densities of TEM wave packet', *Physica*, **43**: 77-91.

Hawthorne, E.I. (1953). 'Flow of energy in D-C machines', *AIEE Transactions*, **72**(1): 438-445.

Heald, M.A. (1984). 'Electric fields and charges in elementary circuits', *American Journal of Physics*, **52**(6): 522-526.

Heald, M.A. (1988). 'Energy flow in circuits with Faraday emf', *American Journal of Physics*, **56**(6): 540-547.

Heaviside, O. (1912). *Electromagnetic Theory*, 3 vols. The Electrician Publishing Company (reprint, 1971).

Hehl, F.W. and Y.N. Obukhov (2003). *Foundations of Classical Electrodynamics. Charge, Flux, and Metric*. Springer Science.

Heimann, P.M. (1974). 'Conversion of forces and conservation of energy', *Centaurus*, **18**(2): 147-161.

Helmholtz, H. (1872). 'On the theory of electrodynamics', *The London, Edinburgh, and Dublin Philosophical Magazine and Journal of Science*, **44**(296): 530-537.

Hendry, J. (1984). *The Creation of Quantum Mechanics and the Bohr-Pauli Dialogue*. Dordrecht: D. Reidel Publishing Co.

Hensel, S. (1994). 'Looking back [Ernst Berg ideas on Heavyside's calculus]', *IEEE Potentials*, **13**(3): 57-60.

Hentschel, K. (2006). 'Light quanta: The maturing of a concept by the stepwise accretion of meaning', *Physics and Philosophy*, 1-20.

Hentschel, K. (2015). 'Die allmähliche Herausbildung des Konzepts Lichtquanten', *Berichte zur Wissenschaftsgeschichte*, **38**(2): 121-139.

Heras, J.A. (1994). 'Jefimenko's formulas with magnetic monopoles and Liénard-Weichert fields of a dual-charged particle', *American Journal of Physics*, **62**(6): 525-531.

Heras, J.A. (2006). 'Instantaneous fields in classical electrodynamics', arXiv: physics/0701007v1 [physics.class-ph] 30 Dec 2006, p. 9.

Heras, J.A. (2006). 'Comment on Helmholtz theorem and the v-gauge in the problem of superluminal and instantaneous signals in classical electrodynamics' by A. Chubykalo *et al.*, *Foundations of Physics Letters*, **19**(6): 579-590.

Heras, J.A. (2008a). Author's response, *American Journal of Physics*, **76**(2): 101-102.

Heras, J.A. (2008b). 'The exact relation between the displacement current and the conduction current: Comment on a paper by Griffiths and Heald', arXiv:0812.4842v1 [physics.class-ph] 28 Dec. 2008.

Heras, J.A. (2009). 'How to obtain the covariant form of Maxwell's equations from the continuity equation', arXiv:0912.5041v1 [physics.class-ph] 26 Dec. 2009, p. 14.

Heras, J.A. (2010a). 'The equivalence principle and the correct form of writing Maxwell's equations', arXiv:1012.1067v1 [physics.class-ph] 6 Dec. 2010, p. 13.

Heras, J.A. (2010b). 'Can the Lorentz-gauge potentials be considered physical quantities?', arXiv:1012.1063v1 [physics.class-ph] 6 Dec. 2010, p. 12.

Heras, J.A. (2011). 'A formal interpretation of the displacement current and the instantaneous formulation of Maxwell's equations', *American Journal of Physics*, **79**(4): 409-416.

Heras, J.A. (2016). 'An axiomatic approach to Maxwell's equations', arXiv:1608.00659v1 [physics.class-ph] 2 August 2016.

Heras, J.A. (2017). 'Alternatives routes to the retarded potentials'. *European Journal of Physics*, **38**(5): 055203. 13 p.

Hering, C. (1923). 'Electromagnetic forces: A search for more rational fundamentals: A proposed revision of the laws', *Midwinter Convention of the American Institution of Electrical Engineers*, February 14-17: 139-154.

Hernandes, J.A. and A.K.T. Assis (2003). 'The potential, electric field and surface charges for a resistive long straight strip carrying a steady current', *American Journal of Physics*, **71**(9): 936-942.

Herrera, R.S. and P. Salmeron (2007). 'Instantaneous reactive power theory: A comparative evaluation of different formulations', *IEEE Transactions on Power Delivery*, **22**(1): 595-604.

Herrmann, F. and G.B. Schmid (1986). 'The Poynting vector field and the energy flow within a transformer', *American Journal of Physics*, **54**(6): 528-531.

Hesse, M.B. (1955). 'Action at a distance in classical physics', *Isis*, **46**(4): 337-353.

Hesse, M.B. (1959). *Review: Concepts of Force: A Study in the Foundations of Dynamics* by Max Jammer. *The British Journal for the Philosophy of Science*, **10**(37): 69-73.

Hesse, M.B. (1961). *Forces and Fields: The Conception of Action at a Distance in the History of Physics*. London: Thomas Nelson and Sons Ltd.

Hesse, M.B. (1964). 'Resource letter PhM-1 on philosophical foundations of classical mechanics', *American Journal of Physics*, **32**(12): 905-911.

Hestenes, D. (1967). 'Real spinor fields', *Journal of Mathematical Physics*, **8**(4): 798-808.

Hestenes, D. (1971). 'Vectors, spinors, and complex numbers in classical and quantum physics', *American Journal of Physics*, **39**(9): 1013-1027.

Hestenes, D. (1982). 'Space-time structure of weak and electromagnetic interactions', *Foundations of Physics*, **12**(2): 153-168.

Hestenes, D. (1990). 'The Zitterbewegung interpretation of quantum mechanics', *Foundations of Physics*, **20** (10): 1213-1232.

Hestenes, D. (1993). 'Differential forms in geometric algebra'. *In:* F. Brackx *et al.* (Eds.). *Clifford Algebras and their Applications in Mathematical Physics*. Dordrecht/Boston: Kluwer, 269-285.

Hestenes, D. (2002a). *Classical Foundations for Classical Mechanics*. New York: Kluwer Academic Publishers.

Hestenes, D. (2002b). *New Foundations for Classical Mechanics*. New York: Kluwer Academic Publishers.

Hestenes, D. (2003). 'Spacetime physics with geometric algebra', *American Journal of Physics*, **71**(7): 691-714.

Hestenes, D. (2014). 'Tutorial on Geometric Calculus', *Advances in Applied Clifford Algebras*, **24**(2): 257-273.

Hestenes, D. (2015). 'Electrodynamics'. In: *Space-Time Algebra*, 25-33. Springer.

Hestenes, D. (2017). 'The genesis of geometric algebra: A personal retrospective', *Advances in Applied Clifford Algebras*, **27**(1): 351-379.

Hestenes, D. (2019). 'Zitterbewegung structure in electrons and photons'. arXiv:1910.11085v1 [physics.gen-ph] 2 Oct. 2019.

Hestenes, D. and R. Gürtler (1971). 'Local observables in Quantum Theory', *American Journal of Physics*, **39**(9): 1028-1038.

Higgins, T.J. (1949). 'History of the operational calculus as used in electric circuit analysis', *Electrical Engineering* [IEEE], **68**(1): 42-45.

Hildebrand, D. (2013). *Foundations of Geometric Algebra Computing*. Berlin: Springer.

Hill, S.E. (2010). 'Rephrasing Faraday's law', *The Physics Teacher*, **48**: 410-412.

Hill, S.E. (2011). 'Reanalyzing the Ampère-Maxwell Law', *The Physics Teacher*, **49**: 343-345.

Hillion, P. and H.F. Harmuth (1991). 'Remark on Harmuth's 'Correction of Maxwell's equations for signals. I' (and response)', *IEEE Transactions on Electromagnetic Compatibility*, **33**(2): 144-145. https://doi.org/10.1109/15.78352.

Hines, C.O. (1952). 'Electromagnetic energy density and flux', *Canadian Journal of Physics*, **30**(2): 123-129.

Hirosige, T. (1969). 'Origins of Lorentz's theory of electrons and the concept of the electromagnetic field', *Historical Studies in the Physical Sciences*, **1**: 151-209.

Hladik, J. (1999). *Spinors in Physics*. Berlin: Springer.

Hochrainer, A. (1957). *Symmetrische Komponenten in Drehstromsysteme.* Springer.

Hong, S. (1995). 'Forging scientific electrical engineering: John Ambrose Fleming and the Ferranti effect', *Isis*, **86**(1): 30-51.

Horn, M.E. (2006). 'Quaternions and Geometric Algebra'. *arXiv:0709.2238* [physics.ed-ph].

Hoseman, G. (1988). *Elektrische Energietechnik.* vol. 3: *Netze.* Hütte, Taschenbücher der Technik, Berlin: Springer.

Hoseman, G. and W. Boeck (1983). *Grundlagen der elektrischen Energietechnik.* Berlin: Springer.

Hospitalier, E. (1885). 'Unity of definition, notation, and symbols used in electricity', *Journal of the Society of Telegraph Engineers & Electricians,* 14: 167-168.

Hossenfelder, S. (2018). *Lost in Math. How Beauty Leads Physics Astray.* New York: Basic Books.

Hughes, E. (1954). *Fundamentals of Electrical Engineering.* London: Longmans Green and Co.

Hunt, B.J. (1984). *The Maxwellians*, Ph.D. thesis, The John Hopkins University, Ph.D. No. 8501649.

Idelchik, V.I. (1989). *Elektricheskie sistema i seti.* Moscow: Energoizdat.

IEEE Guide for Maintenance Methods on Energized Power Lines (2003). In: *IEEE*

Standard 516-2003 (revision of IEEE Std 516–1995) 1–119. https://doi. org/10.1109/IEEESTD.2003.94383.

IEEE Working Group (1996). 'Practical definitions for powers in systems with non-sinusoidal waveforms and unbalanced loads: A discussion', *IEEE Transactions on Power Delivery*, **11**(1): 79-101.

Ilić, M. and J. Zaborszky (2000). *Dynamic and Control of Large Electric Power Systems*. New York: Wiley.

Iliovici, A.D. (1925). 'Définition et mesure de la puissance et de l'énergie réactive', *Bulletin de la Société française des électriciens* (4th series) 5: 931-954.

Iliovici, A.D. (1931). 'Electric Measuring and protecting device', United States Patent Office Patented July 28, 1931, 1, 816,778. Application filled December 21, 1927. 7 p.

Ingram, W.H. (1976). 'Note on the von Helmholtz impedance and its matric generalization', *International Journal of Mathematical Education in Science and Technology*, **7**(2): 217-220.

Ivezić, T. (2005a). 'Axiomatic geometric formulation of electromagnetism with only one axiom: The field equation for the bivector field F with an explanation for the Trouton-Noble experiment', *Foundations of Physics Letters*, **18**(5): 401-429.

Ivezić, T. (2005b). 'The difference between the standard and the Lorentz transformations of the electric and the magnetic fields. Application to motional EMF', *Foundations of Physics Letters*, **18**(4): 301-324.

Ivezić, T. (2005c). 'The proof that Maxwell equations with the 3D E and B are not covariant upon the Lorentz transformations but upon the Standard Transformations: The new Lorentz invariant field equations', *Foundations of Physics Letters*, **35**(9): 1585-1615.

Ivezić, T. (2006). 'The 4D geometric quantities versus the usual 3D quantities. The resolution of Jackson's paradox'. arXiv: physics/0602105 v1 15 Feb. 2006. Pp 1-13.

Ivrlač, M.T. and J.A. Nossek (2010). 'Toward a circuit theory in communication', *IEEE Transactions on Circuits and Systems I: Regular Papers*, **57**(7): 1663-1683.

Ivrlač, M.T. and J.A. Nossek (2011). 'On the association of power-flow with circuit terminals', *2011 International Conference on Electromagnetics in Advanced Applications*, Torino, 2011. 869-872. https://doi.org/10.1109/ICEAA.2011.6046457.

Jackson, D.C. (1896). *Alternating Currents and Alternating Current Machinery*. New York: The Macmillan Company.

Jackson, J.D. (1996). 'Surface charges on circuit wires and resistors play three roles', *American Journal of Physics*, **64**(7): 855-870.

Jackson, J.D. (2006). 'How an antenna launches its input power into radiation: The pattern of the Poynting vector at and near an antenna', *American Journal of Physics*, **74**(4): 280-288.

Jackson, J.D. and L.B. Okun (2001). 'Historical roots of gauge invariance' *Review of Modern Physics*, 73: 663-680.

Jaeger, J.C. (1940). 'The Laplace Transformation Method in Elementary Circuit Theory', *The Mathematical Gazette*, **24**(258): 42-50.

Jäger, K. and F. Heilbronner (2010). *Lexikon der Elektrotechniker*. Berlin: VDE Verlag.

Jammer, M. (1957). *Concepts of Force. A Study in the Foundations of Dynamics*. Cambridge: Harvard University Press.

Jammer, M. (1966). *The Conceptual Development of Quantum Mechanics*. New York: McGraw-Hill Book Co.

Jammer, M. (2006). *Concepts of Simultaneity: From Antiquity to Einstein and Beyond*. Baltimore: The John Hopkins University Press.

Jancewicz, B. (1988). *Multivectors and Clifford Algebra in Electrodynamics*. Singapore: World Scientific.

Janet, P. (1897). 'Sur une application des imaginaires au calcul des courants alternatifs', *L'Éclairage Électrique*, **13**(51): 529-531.

Janet, P. (1905). *Leçons d'Électrotechnique Générale*, vol. 2. Paris: Gauthier-Villars.

Janet, P. (1918). 'Sur l'application des imaginaires au calcul des courants alternatifs', *Revue d'Électricité*, **14**(6): 267-268.

Janssen, M. (2002). 'The Einstein-Besso manuscript: A glimpse behind the curtain of the wizard', Freshman Colloquium *"Introduction to the Arts and Science"*, University of Minnesota. 16 p.

Janssen, M. and M. Mecklenburg (2006). 'From classical to relativistic mechanics: Electromagnetic models of the electron'. In: *Interactions: Mathematics, Physics and Philosophy, 1860–1930*. (Ed.). V.F. Hendricks, K.F. Jorgensen, J. Lützen and S.A. Pedersen. *Boston Studies in the Philosophy and History of Science*, 251: 65-134. Springer.

Jarem, J.M. and P.P. Banerjee (1999). 'Application of the complex Poynting theorem to diffraction gratings', *Journal of the Optical Society of America A*, **16**(5): 1097-1107.

Jefimenko, O.D. (1962). 'Demonstration of the electric fields of current-carrying conductors', *American Journal of Physics*, **30**(1): 19-21.

Jefimenko, O.D. (1966a). *Electricity and Magnetism: An Introduction to the Theory of Electric and Magnetic Fields*. New York: Appleton-Century-Crofts.

Jefimenko, O.D. (1966b). 'Is magnetic field due to an electric current a relativistic effect?' *European Journal of Physics*, **17**(4): 180-182.

Jefimenko, O.D. (2004). 'Presenting electromagnetic theory in accordance with the principle of causality', *European Journal of Physics*, **25**(2): 287-296.

Jefimenko, O.D. (2008). 'Causal equations for electric and magnetic fields and Maxwell's equations: Comment on a paper by Heras', *American Journal of Physics*, **76**(2): 101.

Jeffreys, H. [1927] (1964). *Operational Methods in Mathematical Physics*. New York: Stechert-Hafner Service Agency.

Jeffries, C. (1992). 'A new conservation law for classical electrodynamics', *SIAM Review*, **34**(3): 386-405.

Jeffries, C. (1994). 'Response to a commentary by F.N. Robinson', *SIAM Review*, **36**(4): 638-641.

Jeltsema, D. (2015). 'Time-varying phasors and their application to power analysis'. In: *Mathematical Control Theory I: Nonlinear and Hybrid Control Systems*, (Ed.). M. Kanat Camlibel, A. Agung Julius, Ramkrishna Pasumarthy, Jacquelien M.A. Scherpen, 51-72, *Lecture Notes in Control and Information Sciences*, 461. Springer.

Jeltsema, D. (2016). 'Budeanu's concept of reactive and distortion power revisited', *Przeglad Elektrotechniczny*, **92**(4): 68-73.

Jeltsema, D. and G. Kaiser (2016). 'Active and reactive energy balance equations in active and reactive time', *Proceedings of the 10th International Conference on Compatibility, Power Electronics and Power Engineering (CPE-POWERENG)*, Bydgoszcz, Poland. 29 June-01 July, 2016, 21-26.

Jeltsema, D., J. van der Woude and M. Hartman (2014). 'A novel time-domain perspective of the CPC power theory: Single-phase systems', arXiv:1403.7842v2 [math.OC] 25 Apr. 2014.

Jerison, D., I.M. Singer and D.W. Stroock (Eds.). (1997). *The Legacy of Norbert Wiener: A Centennial Symposium in Honor of the 100th Anniversary of Norbert Wiener's Birth*, symposium held in Cambridge, Massachusetts in October 1994, *Proceedings of Symposia in Pure Mathematics* 60. Providence, RI: American Mathematical Society.

Jiménez, J.L., I. Campos and N. Aquino (2008). 'Exact electromagnetic fields produced by a finite wire with constant current', *European Journal of Physics*, **29**(1): 163-175.

Johannson, I. (2009). 'Mathematical vectors and physical vectors', *Dialectica*, **63**(4): 433-447.

Johnson, D. (2018). *The Photoelectric Effect, Wave-Particle Duality, and Atomic Structure*, a thesis, Perth, Australia, 14 December 2018. viXra:1812.0276. 54 p.

Johnson, D. (2019). 'The spin torus energy model and electricity', *Open Journal of Applied Sciences*, 9: 451-479.

Johnson, J.A. (1933). 'Operating aspects of reactive power', *Transactions of the American Institute of Electrical Engineers*, **52**(3):752-757.

Joot, P. (2009). 'Energy and momentum for complex electric and magnetic field phasors', originally appeared at http://sites.google.com/site/peeeterjoot/math2009/complexFieldEnergy.pdf.

Joot, P. (2015). *Exploring Physics with Geometric Algebra*. © 2014.

Joot, P. (2017). *Electromagnetic Engineering with Geometric Algebra. A Modern Approach to Electromagnetism*. © 2017.

Joot, P. (2018). *Geometric Algebra for Electrical Engineers: Electromagnetism Applications*, April 2018, version V0.1.2.

Joot, P. (2019). *Geometric Algebra for Electrical Engineers: Multivector Electromagnetism*, May 2019, version V0.1.15–6, Monee, IL, 20 January 2020. ISBN 9781987598971.

Josephs, H.J. (1963). 'Postscript to the work of Heaviside', *Journal of the Institution of Electrical Engineers*, **9**(108): 511-512.

Jungnickel, C. and R. McCormach (1986). *Intellectual Mastery of Nature: Theoretical Physics from Ohm to Einstein*, vol. 1, *The Torch of Mathematics 1800-1870*; vol. 2, *The Now Mighty Theoretical Physics, 1870-1925*. Chicago: The University of Chicago Press.

Kafka, H. (1926). *Die ebene Vektorrechnung*. Leipzig: Verlag und Druck von B.G. Teubner.

Kaiser, G. (2003). 'Helicity, polarization, and Riemann-Silberstein vortices', *Journal of Optics A: Pure and Applied Optics* 6. 10.1088/1464-4258/6/5/018.

Kaiser, G. (2004). 'Helicity, polarization, and Riemann-Silberstein vortices', arXiv: math-ph/0309010v2 16 Feb. 2004.

Kaiser, G. (2011). 'Electromagnetic inertia, reactive energy, and energy flow velocity', arXiv:1105.4834v1 [physics. Class-ph] 24 May 2011. Pp 1-20. .

Kaiser, G. (2012). 'The reactive energy of transient EM fields'. arXiv:1201.6575v2 [math-ph] 16 Jul. 2012.

Kaiser, G. (2015). 'Conservation of reactive EM energy in reactive time'. arXiv:1501.01005v1 [physics. optics] 5 Jan. 2015: 1-2, *IEEE Int. Symposium on Antennas & Propagation & USNC/URSI National Radio Science Meeting*: 1704-1705.

Kaiser, G. (2016). 'Completing the complex Poynting theorem: Conservation of reactive energy in reactive time'. arXiv: 1412.3850v2 [math-ph] 25 Jan. 2016. Pp 1-21.

Kalantarow, P.L. and L.R. Neumann (1959). *Theoretische Grundlagen der Elektrotechnik*. Berlin: VEB Verlag Technik.

Kanarev, Ph.M. (2000). 'Model of the electron', *Apeiron: A Journal for Ancient Philosophy and Science*, **7**(3-4): 184-193.

Kanwal, R.P. (1983). *Generalized Functions: Theory and Technique*. New York: Academic Press.

Kapp, G. (1894). *Electric Transmission of Energy and its Transformation, Subdivision, and Distribution*. New York: D. Van Nostrand Company.

Karapetoff, V. (1912). *The Electric Circuit*. New York: McGraw-Hill Book Co.

Karlsson, R. (2005). *Theory and Applications of Tri-axial Electromagnetic Field Measurements*, Ph.D. thesis presented at Uppsala University, Sweden.

Karugaba, S. and O. Ojo (2009). 'Determination of active and reactive powers in multiphase machines', IEEE 978-4244-2893-9/09 – IEEE

Energy Conversion Congress and Expo (ECCE), San Jose, CA., September 20–24, 2009, 617-624.

Katz, E. (1965). 'Concerning the number of independent variables of the classical electromagnetic field', *American Journal of Physics*, **33**(4): 306-312.

Kennelly, A.E. (1893). 'Vectors and Quaternions', *Nature* **48**(1249): 540-541.

Kennelly, A.E. (1910). 'Vector power in alternating-current circuits', a paper presented at the 27th Annual Convention of the American Institute of Electrical Engineers, Jefferson, N.H., June 29: 1233-1267 and Discussions: 1268-1280.

Kennelly, A.E. and E. Velander (1919). 'A rectangular-component two-dimensional alternating-current potentiometer', *Journal of The Franklin Institute*, **188**(1123): 1-26.

Kennelly, A.E. (1928). *Electric Lines and Nets*. McGraw-Hill Book Co., Inc. New York.

Kholmetskii, A.L. (2004). 'Remarks on momentum and energy flux of non-radiating electromagnetic field', *Annales de la Fondation Louis de Broglie*, **20**(3): 549-581.

Kholmetskii, A.L., O. Missevitch and T. Yarman (2011). 'Continuity equations for bound electromagnetic field and the electromagnetic energy–momentum tensor', *Physica Scripta*, **83**(5): 055406. https://doi.org/10.1088/0031-8949/83/05/055406.

Kholmetskii, A.L., O. Missevitch and T. Yarman (2016). 'Poynting Theorem, relativistic transformation of total energy-momentum and electromagnetic energy-momentum tensor', *Foundations of Physics*, **46**(2): 236-261.

Kibble, T.W.B. (2015). 'Spontaneous symmetry breaking in gauge theories', *Philosophical Transactions of the Royal Society A: Mathematical, Physical and Engineering Science*, **373**(2932): 1-12.

Kinsler, P., A. Favaro and M.W. McCall (2009). 'Four Poynting theorems', *European Journal of Physics*, **30**(5): 983-993.

Klein, W. (1981). 'Experimental "paradox" in electrodynamics', *American Journal of Physics*, **49**(6): 603-604.

Kleiner, I, (1988). 'Thinking the unthinkable: The story of complex numbers (with a moral)', *The Mathematics Teacher*, **81**(7): 583-592.

Kleiner, I. (1989). 'Evolution of the function concept: A brief survey', *The College Mathematics Journal*, **20**(4): 282-300.

Kline, M. (1962). *Electromagnetic Theory and Geometrical Optics. Scholar's Choice Edition*, research sponsored by the Electronic Research Directorate of the Air Force Cambridge Research Laboratories, Office of Aerospace Research. Contract No AF 19 (604) 5238. Bedford, MA: USAF.

Kline, R.R. (1983). *Charles P. Steinmetz and the Development of Electrical*

Engineering Science, Ph.D. dissertation, 83-23061, The University of Wisconsin-Madison.

Kobe, D.H. (1982). 'Energy flux vector for the electromagnetic field and gauge invariance', *American Journal of Physics*, **50**(12): 1162-1164.

Kobe, D.H. (2013). 'Noether's theorem and the work-energy theorem for a charged particle in an electromagnetic field', *American Journal of Physics*, **81**(3): 186-189.

Koettnitz, H. and H. Pundt (1968). *Berechnung elektrischer Energieversorgungsnetze*, vol. I, *Mathematische Grundlagen und Netzparameter*. Leipzig: *VEB Deutscher Verlag für Grundstoffindustrie*.

Kojevnikov, A. (1999). 'Freedom, collectivism, and quantum particles: Social metaphors in quantum physics', *Historical Studies in the Physical and Biological Sciences*, **29**(2): 295-331.

Konopinski, E.J. (1978). 'What the electromagnetic vector potential describes', *American Journal of Physics*, **46**(5): 499-502.

Konopinski, E.J. (1981). *Electromagnetic Fields and Relativistic Particles*. New York: McGraw-Hill Book Co.

Koppelman, E. (1971). 'The calculus of operations and the rise of abstract algebra', *Archive for History of Exact Sciences*, **8**(3): 155-242.

Kosso, P. (2000). 'The epistemology of spontaneously broken symmetries', *Synthese*, **122**(3): 359-376.

Kouznetsov, B. and A. Frenk (1974). 'La correspondance Einstein-Besso', *Revue d'histoire des sciences*, **27**(1): 77-82.

Kovacs, K.P. (1960). 'Symmetrische Komponenten der Momentanwerte, oder Vektoren der elektrischen Größen', *Archiv für Elektrotechnik*, **45**(2): 99-117.

Kox, A.J. (2005). 'H.A. Lorentz, Lectures on Electron Theory, 1st ed. (1909)'. In: *Landmark Writings in Western Mathematics, 1640–1940*, (Ed.). I. Grattan-Guinness, 778-783. Elsevier.

Knott, C.G. (1893). 'Vectors and quaternions', *Nature*, **48**(1233): 148-149.

Knowlton, A.E. (1933). 'Reactive power concepts in need of clarification', *Transactions of the American Institute of Electrical Engineers*, **52**(3): 744-747: Discussions: 779-801.

Kracklauer, A.F. (2013). 'Photon statistic: Math versus mysticism', *Proceedings of the Society of Photo-Optical Instrumentation Engineers (SPIE) 8832, The Nature of Light: What are Photons?* V, 88320E.

Kracklauer, A.F. (2015). 'What is a photon?' *Proceedings of the Society of Photo-Optical Instrumentation Engineers (SPIE) 9570, The Nature of Light: What are Photons?* VI, 957008: 1-13.

Kragh, H. (1992). 'Ludwig Lorenz and the early theory of long-distance telephony', *Centaurus*, **35**(3): 305-324.

Kragh, H. (1999. *Quantum Generations. A History of Physics in the Twentieth Century*. Princeton, NJ: Princeton University Press.

Kragh, H. (2018a). 'Ludwig Lorenz and his non-Maxwellian electrical theory of light', *Physics in Perspective*, 20: 221-253.

Kragh, H. (2018b). 'Ludwig Lorenz (1867) on Light and Electricity'. ArXiv, March 16, 2018. Pp 1-8. http://arxiv.org/abs/1803.06371.

Krajewski, R.A. (1990). 'A formal aspect of the definition of power', *Measurement*, **8**(2): 77-83.

Kraus, J.D. and K.R. Carver (1973). *Electromagnetics*. Hightstown, NJ: McGraw-Hill.

Krauss, F. (2010). 'Introduction to particle physics. Lecture 9: Gauge invariance' (power point presentation), Institute for Particle Physics Phenomenology (IPPP), Durham University. http://www.ippp.dur.ac.uk/~krauss/Lectures/IntroToParticlePhysics/2010/Lecture9.pdf.

Kuhn, T.S. (1970). *The Structure of Scientific Revolutions*, 2nd ed., Chicago: The University of Chicago Press.

Kuipers, J.B. (2002). *Quaternions and Rotation Sequences*. Princeton, NJ: Princeton University Press.

Kukačka, L., J. Kraus, M. Kolář, P. Dupuis and G. Zissis (2016). 'A review of AC power theories under stationary and non-stationary, clean and distorted conditions', *IET Generation, Transmission & Distribution*, **10**(1): 221-231.

Kullstam, P.A. (1991). 'Heaviside's operational calculus: Oliver's revenge', *IEEE Transactions on Education*, **34**(2): 155-156.

Kullstam, P.A. (1992). 'Heaviside's operational calculus applied to electrical circuit problems', *IEEE Transactions on Education*, **35**(4): 266-277.

Kulsrud, R.M. (1960). 'Plasma Physics', *American Scientist*, **48**(4): 581-598.

Küpfmüller, K. (1939). *Einführung in die theoretische Elektrotechnik*, 2nd ed. Berlin: Verlag von Julius Springer.

Küpfmüller, K. (1959). *Einführung in die theoretische Elektrotechnik*, 6th ed. Berlin: Springer.

Küpfmüller, K., W. Mathis and A. Reibiger (2013). *Theoretische Elektrotechnik: Eine Einführung*, 19th ed. Berlin: Springer.

Kwong, C.P. (2009). 'The mystery of square root of minus one in quantum mechanics, and its demystification', arXiv:0912.3996v1 [physics.gen.ph] 20 Dec. 2009, 9 p.

Lacava, F. (2016). *Classical Electrodynamics: From Image Charges to the Photon Mass and Magnetic Monopoles*, Undergraduate Lecture Notes in Physics. Springer International Publishing.

Lachièze-Rey, M. (2009). 'Spin and Clifford Algebras: An Introduction', *Advances in Applied Clifford Algebras*, **19**(3): 687-720.

Lai, C.S. (1981). 'Alternative choice for the energy flow vector of the electromagnetic field', *American Journal of Physics*, **49**(9): 841-843.

Lakatos, I. (1976). *Proofs and Refutations. The Logic of Mathematical Discovery*. Cambridge: Cambridge University Press.

Landau, L.D. and E.M. Lifshitz (1975). *The Classical Theory of Fields*. Oxford: Pergamon Press.

Langacker, P. (2017). *Can the Laws of Physics be Unified?* Princeton, NJ: Princeton University Press.

Lasenby, A., C. Doran and E. Arcaute (2004). 'Applications of Geometric Algebra in Electromagnetism, Quantum Theory and Gravity'. In: *Clifford Algebras* (Ed.). R. Abłamowicz, Progress in Mathematical Physics 34. Boston: Birkhäuser.

Lasenby, A.N. (2017). 'Geometric Algebra as a unifying language for physics and engineering and its use in the study of gravity', *Advances in Applied Clifford Algebras*, **27**(1): 733-759.

Lasenby, J., A.N. Lasenby and C.J.L. Doran (2000). 'A unified mathematical language for physics and engineering in the 21st century', *Philosophical Transactions of the Royal Society A: Mathematical, Physical and Engineering Sciences*, **358**(1765): 21-39.

Laughton, M.A. and M.G. Say (1991). *Electrical Engineer's Reference Book*. London: Butterworth-Heinemann.

Law, J. and R. Rennie (Eds.). (2015). *A Dictionary of Physics*, 7th ed. Oxford: Oxford University Press.

LaWhite, N. (1995). *Vector Calculus of Periodic Non-sinusoidal Signals for Decomposition of Power Components in Single and Multiphase Circuits*, Master of Science thesis, Massachusetts Institute of Technology, 1-89.

LaWhite, N. and M.D. Ilic (1997). 'Vector space decomposition of reactive power for periodic non-sinusoidal signals', *IEEE Transactions on Circuits and Systems – I: Fundamental Theory and Applications*, **44**(3): 338-346.

Lazarovici, D. (2018). 'Against Fields', *European Journal for Philosophy of Science*, **8**(2): 145-170.

Le Bellac, M. (2014). *An Introduction to the Quantum World*. Singapore: World Scientific.

Le Bellac, M. and J.-M. Lévy-Leblond (1973). 'Galilean Electro-magnetism', *Il Nuovo Cimento*, **14B**(2): 217-234.

Lehrman, Robert L. (1973). 'Energy is not the ability to do work', *The Physics Teacher*, 11: 15. https://doi.org/10.1119/1.2349846.

Lenin, V.I. [1909] (1972). *Materialism and Empirio-criticism: Critical Comments on a Reactionary Philosophy. Lenin Collected Works*, vol. 14, translated by Abraham Fineberg. Moscow: Progress Publishers, 1972, 17-362.

Leus, V.A., R.T. Smith and S. Maher (2013). 'The physical entity of vector potential in electromagnetism', *Applied Physics Research*, **5**(4): 56-68.

Lev-Ari, H. and A.M. Stankovic (2003). 'Hilbert space techniques for modeling and compensation of reactive power in energy processing systems', *IEEE Transactions on Circuits and Systems I: Fundamental Theory and Applications*, **50**(4): 540-556.

Lev-Ari, H. and A.M. Stankovic (2006). 'A decomposition of apparent power in polyphase unbalanced networks in nonsinusoidal operation', *IEEE Transactions on Power Systems*, **21**(1): 438-440.

Lev-Ari, H. and A.M. Stankovic (2009a). 'Instantaneous power quantities in polyphase systems: A geometric algebra approach', *2009 IEEE Energy Conversion Congress and Exposition (ECCE)*, San Diego, CA. IEEE 978-1-4244-2893-9: 592-596.

Lev-Ari, H. and A.M. Stankovic (2009b). 'A geometric algebra approach to decomposition of apparent power in general polyphase networks', *41st North American Power Symposium (NAPS 2009)*, Starkville, MS, 376-381.

Lev-Ari, H., A.M. Stankovic and S.J. Ceballos (2005). 'An orthogonal decomposition of apparent power with application to an industrial load', *XV Power Systems Computation Conference (PSCC)*, Liège, 22-26 August 2005, Session 22, Paper No. 5: 1-7.

Lewis, G.N. (1910). 'On four-dimensional vector analysis, and its applications in electrical theory', *Proceedings of the American Academy of Arts and Sciences*, **46**(7): 165-181.

Li, Hongbo (2008). *Invariant Algebras and Geometric Reasoning*. River Edge, NJ: World Scientific.

Lindell, I.V. (1983). 'Complex vector algebra in electromagnetics', *International Journal of Electrical Engineering & Education*, **20**(1): 33-46.

Lindell, I.V. (1992). *Methods for Electromagnetic Field Analysis*. Oxford University Press.

Lindell, I.V. (2000). 'Heaviside operational rules applicable to electromagnetic problems', *Progress in Electromagnetics Research*, 26: 293-331.

Lindell, I.V. (2004). *Differential Forms in Electromagnetism*. Wiley-Interscience.

Lindell, I.V. (2005). Electromagnetic wave equation in differential-form representation, *Progress in Electromagnetics Research*, 54: 321-333.

Lipschitz, R. (1880). 'Principes d'un calcul algébrique qui contient comme espèces particulières le calcul des quantités imaginaires et des quaternions', *Comptes rendus de l'Académie des Sciences Paris*, 91:619-621; 660-664.

Liu, Y. and G.T. Heydt (2005). 'Power system harmonics and power quality indices', *Electric Power Components and Systems*, **33**(8): 833-844.

Lomonosov, V.Iu. and K.M. Polivanov. *Elektrotehnika* (in Russian). Moscow: *Gosudarstvennoe Energeticheskoe Izdatelstvo*.

Lorentz, H.A. (1904). *The Theory of Electrons*, 2nd ed. Leipzig: Teubner.

Lorrain, P. (1982). 'Alternative choice for the energy flow vector of the electromagnetic field', *American Journal of Physics*, **50**(6): 492.

Lorrain, P. (1984). 'The Poynting vector in a transformer', *American Journal of Physics*, **52**(11): 987-988.

Lorrain, P., D.R. Corson and F. Lorrain (1996). *Electromagnetic Fields and Waves Including Electric Circuits*. New York: W.H. Freeman and Co.

Loudon, R., L. Allen and D.F. Nelson (1997). 'Propagation of electromagnetic energy and momentum through an absorbing dielectric', *Physical Review E*, **55**(1): 1071-1085.

Lounesto, P. (1996). 'Clifford Algebras and Spinor Operators'. In: *Clifford (Geometric) Algebras*, (Ed.). W.E. Baylis, 5-35. Boston: Birkhäuser.

Lounesto, P. (2001). *Clifford Algebras and Spinors*, 2nd ed., London Mathematical Society Lecture Note Series 286. Cambridge, New York: Cambridge University Press.

Lundin, U., B. Bolund and M. Leijon (2007). 'Poynting vector analysis of synchronous generators using field simulations', *IEEE Transactions on Magnetics*, **43**(9): 3601-3606.

Lützen, J. (1979). 'Heaviside's operational calculus and the attempts to rigorize it', *Archive for History of Exact Sciences*, **21**(2): 161-200.

Lützen, J. (1982). *The Prehistory of the Theory of Distributions*. Heidelberg: Springer.

Lyon, W.V. (1920). 'Reactive power and unbalanced circuits', *Electrical World*, **75**(25): 1417-1420.

Lyon, W.V. (1933). 'Reactive power and power factor', *Transactions of the American Institute of Electrical Engineers*, **52**(3): 763-770.

Lyre, H. (2009). 'Gauge symmetry'. In: *Compendium of Quantum Physics: Concepts, Experiments, History and Philosophy*, (Ed.). Greenberger *et al.*, 248-255. Heidelberg: Springer.

Macdonald, A. (2010). *Linear and Geometric Algebra*. ISBN 9781453854938.

Macdonald, A. (2017). 'A survey of geometric algebra and geometric calculus', *Advances in Applied Clifford Algebras*, **27**(1): 853-891.

Macfarlane A. (1893a). 'On the analytical treatment of alternating currents'. In: *Proceedings of the International Electrical Congress*, Chicago, 21–25 August 1893, 24-31; 74-75. New York: American Institute of Electrical Engineers.

Macfarlane A. (1893b). Book review: *Alternating Currents: An Analytical and Graphical Treatment* by F. Bedell and A.C. Crehore, *Physical Review* (Series 1) 1: 68-70.

Macfarlane A. (1897). 'Application of hyperbolic analysis to the discharge of a condenser', a paper presented at the Annual Meeting of the American Institute of Electrical Engineers, New York, May 18th, 1897, 163-174; discussions: 175-181; 250-252.

Macfarlane, A. (1899). 'The fundamental principles of algebra', *Science*, **10**(246): 345-364.

Macfarlane, A. (1900). 'Hyperbolic quaternions', *Proceedings of the Royal Society at Edinburgh 1899-1900 Session*, 23: 169-181.

Machowski, J., J.W. Bialek and J.R. Bumby (1997). *Power System Dynamics and Stability*. Chichester: Wiley.

Magnus, W. and W. Schoenmaker (2002). *Quantum Transport in Submicron Devices: A Theoretical Introduction.* Berlin: Springer.

Malengret, M. and C.T. Gaunt (2011). 'General theory of instantaneous power for multi-phase systems with distortion, unbalance and direct current components', *Electric Power Systems Research,* **81**(10): 1897-1904.

Mancosu, P. (2008). *The Philosophy of Mathematical Practice.* New York: Oxford University Press.

Mansuripur, M. (2011). *Field, Force, Energy and Momentum in Classical Electrodynamics.* Bentham Science Publishers.

Marcus, A. (1941). 'The electric field associated with a steady current in long cylindrical conductor', *American Journal of Physics,* **9**(4): 225-226.

Marganitz, A. (1992). 'Power measurement of periodic current and voltage by digital signal processing', *European Transactions on Electrical Power,* **2**(2): 117-123.

Markovic, M. and Y. Perriard (2011). 'An analytical solution for the torque and power of a solid-rotor induction motor', *IEEE International Electric Machines & Drives Conference* (IEMDC): 1053-1057.

Marten, W. and W. Mathis (1992). *'Zur algebraischen Lösung von Wechselstromaufgaben* (On the algebraic solution of AC problems)', *Frequenz,* **46**(3-4): 95-101.

Matar, M. and R. Welti (2017). 'Surface charges and J. H. Poynting's disquisitions on energy transfer in electrical circuits, *European Journal of Physics,* **38**(6): 065201. 16 p.

Mathis, W. (1987). *Theorie nicht-linearer Netzwerke.* Berlin: Springer.

Mathis, W. (1994). 'Analysis of power in nonlinear electrical circuits', *Technical Reports of the Technical University of Budapest,* Electrical Engineering Series, *Intern. Journal No. 5,* joint to International Symposium on Theoretical Electrotechnics (TUB-TR-94-EE) 13, Budapest: 53-60.

Mathis, W. and W. Marten (1989). 'A unified concept of electrical power', *Proceedings of the International Symposium on Circuit Systems* (ISCAS '89, Portland, Oregon, 9-11 May): 499-502.

Mathur, R.M. and R.K. Varma (2002). *Thyristor-based FACTS Controllers for Electrical Transmission Systems.* New York: IEEE Press and Wiley-Interscience.

Matzek, M.A. and B.R. Russell (1968). 'On the transverse electric field within a conductor carrying a steady current', *American Journal of Physics,* **36**(10): 905-907.

Maxwell, J.C. (1865). 'A dynamical theory of the electromagnetic field', *Philosophical Transactions of the Royal Society,* London, 155: 459-512.

Maxwell, J.C. (2003). *The Scientific Papers of James Clerk Maxwell.* Mineola, NY: Dover Publications Inc.

Maxwell, N. (2017). *Karl Popper, Science and Enlightenment*. London: UCL Press.

Mayer, Ch. L. (1938). 'Le mouvement helicoïdal des photons', *Fondation pour les Sciences Phys.-Chim.* 16 p.

McDonald, K.T. (2016). 'Momentum in a DC circuit', *K. McDonald's Physics Examples*. Joseph Henry Laboratories, Princeton University. http://www.physics.princeton.edu/~mcdonald/examples/loop.pdf.

McDonald, K.T. [2013] (2020). '729 variants of Poynting's theorem', *K. McDonald's Physics Examples*. Joseph Henry Laboratories, Princeton University. 5 p. http://www.physics.princeton.edu/~mcdonald/examples/variants.pdf.

McDonald, K.T. [2018] (2019a.). 'Alternative forms of the Poynting vector', *K. McDonald's Physics Examples*. Joseph Henry Laboratories, Princeton University. 20 p. http://www.physics.princeton.edu/~mcdonald/examples/poynting_alt.pdf.

McDonald, K.T. [2010] (2019b). 'Charge density in a current-carrying wire', *K. McDonald's Physics Examples*. Joseph Henry Laboratories, Princeton University. http://www.physics.princeton.edu/~mcdonald/examples/wire.pdf.

McGuyer, B.H. (2012). 'Symmetry and voltmeters', *American Journal of Physics*, **80**(2): 101.

McLachlan, N.W. (1938). 'Historical Note on Heaviside's Operational Method', *The Mathematical Gazette*, **22**(250): 255-260.

Mead, C.A. (1997). 'Collective electrodynamics I', *Proceedings of the National Academy of Sciences of the United States of America*, **94**(12): 6013-6018.

Mehra, J. (Ed.). (1973). *The Physicist's Conception of Nature* [Symposium on the Development of the Physicist's Conception of Nature in the Twentieth Century held at the Abdus Salam International Centre for Theoretical Physics in Miramare, Trieste, Italy in 1972]. Dordrecht: D. Reidel Publishing Co.

Menti, A., T. Zacharias and J. Milias-Argitis (2007). 'Geometric algebra: A powerful tool for representing power under non-sinusoidal conditions', *IEEE Transactions on Circuits and Systems I: Fundamental Theory and Applications*, **54**(3): 601-609.

Menti, A., T. Zacharias and J. Milias-Argitis (2010). 'Power components under non-sinusoidal conditions using a power multivector', *2010 International School on Non-sinusoidal Currents and Compensation*, Lagow, Poland: 174-179.

Merino, O. (2006). *'A short history of complex numbers'*. https://pdfs.semanticscholar.org/2791/0efdc62c67e8d6a307441eabe127bb3546c4.pdf?_ga=2.100628696.717863384.1595295965-861632863.1590204482.

Meulenberg, A., W.R. Hudgins and R.F. Penland (2015). 'The Photon: EM fields, electrical potentials, and AC charge', *Proceedings of the Society of*

Photo-Optical Instrumentation Engineers (SPIE) 9570, *The Nature of Light: What are Photons?* VI, 95700C (10 September 2015). 12 p.

Mikusinski, J.G. (1949). 'Sur le calcul opératoire', Časopis pro pěstování matematiky a fysiky. **74**(2): 89-94.

Mikusinski, J.G. (1950). 'Sur les fondements du calcul opératoire', *Studia Mathematica,* **11**(1): 41-70.

Mikusinski, J.G. (1959). *Operational Calculus.* London: Pergamon Press.

Milanez, D.L. and A.E. Emanuel (2003). 'The instantaneous-space-phasor a powerful diagnosis tool', *IEEE Transactions on Instrumentation and Measurement,* **52**(1): 143-148.

Miller, A.I. (1981). 'Unipolar induction: A case study of the interaction between science and technology', *Annals of Science,* **38**(2): 155-189.

Miller, A.I. (1987). *Imagery in Scientific Thought: Creating 20th-Century Physics.* Cambridge: MIT Press.

Miloni, P.W. (2005). *Fast Light, Slow Light and Left-handed Light.* IOP Publishing Ltd. Bristol, UK.

Milton, K.A. and J. Schwinger (2006). *Electromagnetic Radiation: Variational Methods, Waveguides and Accelerators.* Berlin: Springer.

Miyamoto, K. (2008). *Plasma Physics and Controlled Nuclear Fusion.* Berlin: Springer.

Montgomery, C.G., R.H. Dicke and E.M. Purcell (Eds.). (1948). *Principles of Microwave Circuits,* Radiation Laboratory Series 8. New York: McGraw-Hill.

Monticelli, A. (1999). *State Estimation in Electric Power Systems: A Generalized Approach.* Boston: Kluwer Academic Publishers.

Montoya, F.G. (2019). 'Geometric Algebra in non-sinusoidal power systems: A case study for passive compensation', *Symmetry,* **11**(10): 1287. https://doi.org/10.3390/sym11101287.

Montoya, F.G., A. Alcayde, F.M. Arrabal-Campos, and R. Baños (2019a). 'Quadrature current compensation in non-sinusoidal circuits using geometric algebra and evolutionary algorithms', *Energies,* **12**(4): 692. 17 p. https://doi.org/10.3390/en12040692.

Montoya, F.G., R. Baños, A. Alcayde and F.M. Arrabal-Campos (2019b). 'Analysis of power flow under non-sinusoidal conditions in the presence of harmonics and interharmonics using geometric algebra', *International Journal of Electrical Power and Energy Systems,* **111**(October 2019): 486-492.

Montoya, F.G., R. Baños, A. Alcayde, F.M. Arrabal-Campos and E. Viciana (2020). 'Analysis of non-active power in non-sinusoidal circuits using geometric algebra', *International Journal of Electrical Power and Energy Systems,* 116 (March 2020) 105541: 1-9.

Moon, P. and D.E. Spencer (1954a). 'A new electrodynamics', *Journal of the Franklin Institute,* **257**(5): 369-382.

Moon, P. and D.E. Spencer (1954b). 'The Coulomb force and the Ampère force', *Journal of the Franklin Institute*, **257**(4): 305-315.

Moon, P. and D.E. Spencer (1954c). 'Electromagnetism without Magnetism: An historical sketch', *American Journal of Physics*, **22**(3): 120-124.

Moon, P. and D.E. Spencer (1955a). 'A postulational approach to electromagnetism', *Journal of the Franklin Institute*, **259**(4): 293-305.

Moon, P. and D.E. Spencer (1955b). 'On the Ampère force', *Journal of the Franklin Institute*, **259**(4): 295-311.

Moon, P. and D.E. Spencer (1960). *Foundations of Electrodynamics*. New York: D. Van Nostrand Company, Inc.

Moorcroft, D.R. (1968). 'Faraday's Law – Demonstration of a teaser', *American Journal of Physics*, **37**(2): 221.

Moorcroft, D.R. (1969). 'Faraday's Law, Potential and Voltage – Discussion of a teaser', *American Journal of Physics*, **38**(3): 376.

Moore, D.H. (1966). 'Another approach to Mikusinski's validation of Heaviside operational calculus', *IEEE Transactions on Education*, **E-9**(3): 154-165.

Moore, D.H. (1968). 'Algebraic Basis of Heaviside Operational Calculus', *Journal of the Franklin Institute*, **286**(2): 158-161.

Moore, D.H. (1970). 'Complex number algebra as a simple case of Heaviside operational calculus', *Mathematics Magazine*, **43**(5): 269-272.

Moore, D.H. (1971). *Heaviside Operational Calculus. An Elementary Foundation*. New York: American Elsevier.

Moore, D.H. (1995). 'The axiomatization of linear algebra: 1875-1940', *Historia Mathematica*, **22**(3): 262-303.

Morais, J.P., S. Georgiev and W. Sprößig (2014). *Real Quaternionic Calculus Book*. Basel: Springer.

Morando, A.P. (2001). 'A thermodynamic approach to the instantaneous non-active power', *European Transactions on Electrical Power*, **11**(6): 357-364.

Moriyasu, K. (1978). 'Gauge invariance rediscovered', *American Journal of Physics*, **46**(3): 274-278.

Moriyasu, K. (1980). 'Breaking of gauge symmetry: A geometrical view', *American Journal of Physics*, **48**(3): 200-204.

Moriyasu, K. (1982). 'The renaissance of gauge theory', *Contemporary Physics*, **23**(6): 553-581.

Morris, N.A. and D.F. Styer (2012). 'Visualizing Poynting vector energy flow in electric circuits', *American Journal of Physics*, **80**(6): 552-554.

Morrison, M. (2007). 'Spin: All is not what it seems', *Studies in History and Philosophy of Modern Physics*, **38**(3): 528-557.

Morton, N. (1979). 'An introduction to the Poynting vector', *Physics Education*, **14**(5): 301-304.

Mukerji, S.K., S.K. Goel, S. Bhooshan and K.P. Basu (2004). 'Electromagnetic fields theory of electrical machines, Part I: Poynting theorem for electromechanical energy conversion', *International Journal of Electrical Engineering & Education*, **41**(2): 137-145.

Müller, C. (1969). *Foundations of the Mathematical Theory of Electromagnetic Waves*. Springer.

Müller, C. (2012). 'A semiquantitative treatment of surface charges in DC circuits', *American Journal of Physics*, **80**(9): 782-788.

Muralidhar, K. (2011). 'Classical origin of quantum spin', *Apeiron: A Journal for Ancient Philosophy and Science*, **18**(2): 146-160.

Muralidhar, K. (2014). 'Complex vector formalism of harmonic oscillator in geometric algebra: Particle mass, spin and dynamics in complex vector space', *Foundations of Physics*, **44**(3): 266-295.

Muralidhar, K. (2015a). 'Algebra of complex vectors and applications in electromagnetic theory and quantum mechanics', *Mathematics*, **3**(3): 781-842.

Muralidhar, K. (2015b). 'Classical approach to the quantum condition and biaxial spin connection to the Schrödinger equation', *Quantum Studies: Mathematics and Foundations*, **3**(1): 31-39.

Muralidhar, K. (2016a). 'Mass of a charged particle with complex structure in zero-point field', *Progress in Physics* **12**(3): 224-230.

Muralidhar, K. (2016b). 'The structure of the photon in complex vector space', *Progress in Physics*, **12**(3): 291-296.

Muralidhar, K. (2017). 'Theory of stochastic Schrödinger equation in complex vector space', *Foundations of Physics*, **47**(4): 532-552.

Muralidhar, K. (2018). 'The relativistic effects of charged particle with complex structure in zero-point field', *Open Access Journal of Mathematical and Theoretical Physics*, **1**(4): 125-130.

Murnaghan, F.D. (1951). 'What is a spinor?' *The Scientific Monthly*, **72**(4): 223-224.

Nahin, P.J. (1985). 'Oliver Heaviside, fractional operators, and the age of the Earth', *IEEE Transactions on Education*, **28**(2): 94-104.

Nahin, P.J. (1988). *Oliver Heaviside: Sage in Solitude*. New York: IEEE Press.

Nahin, P.J. (1998). *An Imaginary Tale: The Story of $\sqrt{(-1)}$*. Princeton, NJ: Princeton University Press.

Nahin, P.J. (2019). *Transients for Electrical Engineers: Elementary Switched-circuit Analysis in the Time and Laplace Transform Domains (with a touch of MATLAB)*. Springer International.

Nasar, S.A. (1990). *Electric Power Systems*. New York: McGraw-Hill.

Nedelcu, V.N. (1963). 'Die einheitlische Leistungstheorie der unsymmetrischen und mehrwelligen Mehrphasensysteme', *ETZ-Archiv für Elektrotechnik*, **84**(5): 153-157.

Nedelcu, V.N. (1968). *Regimurile de funcţionare ale maşinilor de curent alternativ* (in Romanian). Bucharest: EdituraTehnică.

Needham, T. (1997). *Visual Complex Analysis*. Oxford: Clarendon Press.

Nelson, D.F. (1979). *Electric, Optic, and Acoustic Interactions in Dielectrics.* New York: Wiley.

Nelson, D.F. (1991). 'Momentum, pseudomomentum, and wave momentum: Toward resolving the Minkowski-Abraham controversy', *Physical Review A,* **44**(6): 3985-3996.

Nelson, D.F. (1995). 'Deriving the transmission and reflection coefficients of an optically active medium without using boundary conditions', *Physical Review E,* **51**(6): 6142-6153.

Nelson, D.F. (1996). 'Generalizing the Poynting vector', *Physical Review Letters,* **76**(25): 4713-4716.

Neumann, C. (1893). *Beiträge zu einzelnen Theilen der mathematischen Physik, insbesondere zur Elektrodynamik und Hydrodynamik, Elektrostatik und magnetischen Induktion.* Leipzig: B.G. Teubner.

Neumann, F. (1845). *Die mathematischen Gesetze der inducirten elektrischenströme.* Leipzig: Verlag von Wilhelm Engelmann.

Nichols, H.W. (1917). 'Theory of variable dynamical-electrical systems', *Physical Review,* **10**(2): 171-193.

Nicholson, H.W. (2005). 'What does the voltmeter read?' *American Journal of Physics,* **73**(12): 1194-1196.

Nikouravan, M. (2019). 'A short history of imaginary numbers', *International Journal of Fundamental Physical Sciences,* **9**(1): 1-5.

Noether, E. (1918). 'Invariante Variationsprobleme', *Nachrichten von der Gesellschaft der Wissenschaften zu Göttingen, Mathematisch-Physikalische Klasse,* 1918: 235-257. <http://eudml.org/doc/59024>.

Nossek, J.A. (2011). 'On the relation of circuit theory and signals, systems and communications', *AEU International Journal of Electronics and Communications,* **65**(8): 713-717.

Nossek, J.A. and M.T. Ivrlač (2011). 'On the relation of circuit theory and signals, systems and communications', *2011 IEEE International Symposium of Circuits and Systems (ISCAS),* Rio de Janeiro: 603-604.

Nowomiejski, Z. (1964). 'Moszcznost aktiwnaja, reaktiwnaja i moszcznost izkazenija w elektriczeskich sistemah s periodiczeskimi nie sinusoidalnymi prozessami', *Elektromechanika,* USSR, Russia (6): 657-664.

Nowomiejski, Z. (1967). 'Analyse elektrischer Kreise mit periodischen nichtsinusoidalförmigen Vorgängen', *Wissenschaftliche Zeitschrift der Elektrotechnik,* 8: 244-254.

Nowomiejski, Z. (1981). 'Generalized theory of electric power', *Archiv für Elektrotechnik,* **63**(3): 177-182.

Obukhov, Y.N. and F.W. Hehl (2003). 'Electromagnetic energy-momentum and forces in matter', arXiv: physics/0303097v1[physics.class-ph] 22 Mar: 1-17.

Okun, L.B. (1989). 'The concept of mass', *Physics Today,* **42**(6): 31-36.

Okun, L.B. (2006). 'Formula $E = mc^2$ in the year of physics', *Acta Physica Polonica B*, **37**(4): 1327-1332.

Olariu, S. (1990). 'Quantum interference of relativistic electrons in the presence of magnetic fields', *Physics Letters A*, **144**(6-7): 287-292.

Olariu, S. and I.I. Popescu (1986). 'Electromagnetic interactions in multiconnected spaces', *Physical Review D*, **33**(6): 1701-1708.

Oldham, K.T.S. (2008). *The Doctrine of Description: Gustav Kirchhoff, Classical Physics, and the 'Purpose of All Science' in 19th-Century Germany*. Ph.D. thesis, UMI Number: 3331743. University of California, Berkeley.

Olejniczak, K. and G. Heydt (1991). 'A comparison of alternative transforms for electric power engineering applications', *Proceedings of the 22nd Annual North American Power Symposium*, 15-16 October, 1990: 84-93.

Olive, K.A., *et al.* (Particle Data Group) (2014). 'Review of Particle Physics', *Chinese Physics* C38 (9) 090001. 1676 p.

Olsen, P.T., R.E. Elmquist, W.D Phillips, G.R. Jones and V.E. Bower (1989). 'A measurement of the NBS electrical watt in SI units', *IEEE Transactions on Instrumentation and Measurement*, **38**(2): 238-244.

Olsen, R.G. and P.S. Wong (1992). 'Characteristics of low-frequency electric and magnetic fields in the vicinity of electric power lines', *IEEE Transactions on Power Delivery*, **7**(4): 2046-2055.

O'Neill, J. (1986). 'Formalism, Hamilton and Complex Numbers', *Studies in the History and Philosophy of Science*, **17**(3): 351-372.

O'Rahilly, A. (1965). *Electromagnetic Theory: A Critical Examination of Fundamentals*. New York: Dover.

O'Raifeartaigh, L. (1986). *Group Structures of Gauge Theories*. Cambridge University Press.

Orlich, E.M. (1911). *Die Theorie der Wechselströme*. Leipzig: BG Teubner.

Ortega-Calderon, J.E. (2008). *Modeling and Analysis of Electric Arc Loads Using Harmonic DomainTechniques*, Ph.D. thesis, University of Glasgow.

Özdemir, M. (2018). 'Introduction to hybrid numbers', *Advances in Applied Clifford Algebras*, **28**(1): Article 11.

Pai, M.A. (1979). *Computer Techniques in Power System Analysis*. New Delhi: Tata McGraw-Hill Publishing Company Limited.

Paiva, J.P.S. (2005). *Redes de energia eléctrica: Una análise sistémica*. Lisbon: IST Press.

Palit, B.B. (1980a). 'Einheitliche Untersuchung der elektrischen Maschinen mit Hilfe des Poynting-Vektors und des elektromagnetischen Energieflusses im Luftspaltraum. Teil I: Aufbau der Theorie', *Zeitschrift für angewandte Mathematik und Physik* (ZAMP), 31: 384-399.

Palit, B.B. (1980b). 'Einheitliche Untersuchung der elektrischen Maschinen mit Hilfe des Poynting-Vektors und des elektromagnetischen Energieflusses im Luftspaltraum. Teil II: Anwendung der Theorie', *Zeitschrift für angewandte Mathematik und Physik* (ZAMP), 31: 400-412.

Palit, B.B. (1982). 'On active and reactive power flow in electrical machines: A unified analysis', *International Journal of Electrical Engineering & Education*, **19**(1): 67-78.

Pappas, P.T. (1983). 'The original Ampère force and Biot-Savart and Lorentz forces', *Il Nuovo Cimento* **73B**(2): 189-197.

Parra-Serra, Josep (2009). 'Clifford Algebra and the Didactics of Mathematics', *Advances in Applied Clifford Algebras*, **19**(3): 819-834.

Pasko, M. and M. Maciazek (2012). 'Principles of electrical power control'. In: *Power Theories for Improved Power Quality* (Ed.). G. Benysek and M. Pasko, 13-47. London: Springer.

Patterson, G.W. (1907). 'The use of complex quantities in alternating currents', paper presented at the Chicago meeting of the Physical Society, December 30, 1907 to January 2, 1908, *Physical Review*, **26**(3): 266-271.

Pavella, M. and P.G. Murthy (1994). *Transient Stability of Power Systems*. Chichester: Wiley.

Pawlikowski, G.J. (1967). 'The men responsible for the development of vectors', *The Mathematics Teacher*, **60**(4): 393-396.

Payne, W.T. (1952). 'Elementary spinor theory', *American Journal of Physics*, **20**(5): 253-262.

Pennell, W.O. (1929). 'A generalization of Heaviside's expansion theorem', *The Bell System Technical Journal*, **8**(3): 482-492.

Pepper, M. (2011). 'One-dimensional electron transport in semiconductor nanostructures', *Current Science*, **100**(4): 484-487.

Perwass, C. (2009). *Geometric Algebra with Applications in Engineering*. Heidelberg: Springer.

Peskin, M.E. (2019). *Concepts of Elementary Particle Physics*. Oxford: Oxford University Press.

Peters, P.C. (1982). 'Objections to an alternative energy flow vector', *American Journal of Physics*, **50**(12): 1165-1166.

Peters, P.C. (1984). 'The role of induced emf's in simple circuits', *American Journal of Physics*, **52**(3): 208-211.

Petroianu, A.I. (1969). 'A geometrical approach to the steady state problem of electrical networks', *Revue Roumaine des Sciences Techniques, Série Électrotechnique et Énergétique*, **14**(4): 623-630.

Petroianu, A.I. (2014). 'Mathematical representations of electrical power: Vector or complex number? Neither!', *IEEE Electrical Power and Energy Conference* (EPEC), November 12-14, Calgary, Alberta, Canada: 170-177.

Petroianu, A.I. (2015). 'A geometric algebra reformulation and interpretation of Steinmetz's symbolic method and his power expression in alternating current electrical circuits', *Electrical Engineering*, 97: 175-180.

Petrov, A.E. (2008). 'Tensor method and dual networks in electrical engineering', *Russian Electrical Engineering*, **79**(12): 2-12.

Petrova, S.S. (1987). 'Heaviside and the development of the symbolic method', *Archive for History of Exact Sciences*, **37**(1): 1-23.

Petzold, D.W. (1980). *On Electromagnetic Power Waves and Power Density Components*, Ph.D. dissertation, Marquette University, 1980. ProQuest Dissertations and Theses No. 811862.

Philippow, E. (1988). *Taschenbuch Elektrotechnik, Band 6, Systeme der Elektroenergietechnik*. Berlin: VEB Verlag Technik.

Phillips, M. (1962). 'Classical Electrodynamics'. In: *Principles of Electrodynamics and Relativity/ Prinzipien der Elektrodynamik und Relativitätstheorie*, (Ed.). S. Flügge, *Encyclopedia of Physics/ Handbuch der Physik* Series HDBPHYS, vol. 2/4. Heidelberg: Springer.

Pierce, B. (1881). 'Linear Associative Algebra', *American Journal of Mathematics*, **4**(1): 97-229.

Pierseaux, Y. and G. Rousseaux (2006). 'Les structures fines de l'électromagnétisme et de la relativité restreinte', HAL, Archives-Ouverte, 2006, pp.19. arXiv: physics/0601023v1 [physics. hist-ph] 5 Jan. 2006.

Pietsch, W. (2008). 'Two electrodynamics between plurality and reduction', In: *Reduction and the Special Sciences* (Tilburg, April 10-12, 2008). http://philsci-archive.pitt.edu/archive/00003876/.

Pietsch, W. (2010). 'On conceptual issues in classical electrodynamics: Prospect and problems of an action-at-a-distance interpretation', *Studies in History and Philosophy of Modern Physics*, **41**(1): 67-77.

Pietsch, W. (2011). 'The under-determination debate: How lack of history leads to bad philosophy'. In: *Integrating History and Philosophy of Science: Problems and Prospects* (Ed.). S. Mauskopf and R. Schmaltz, *Boston Studies in the Philosophy of Science*, 263. 83-106.

Pietsch, W. (2012). 'Hidden under-determination: A case study in classical electrodynamics', *International Studies in the Philosophy of Science*, **26**(2): 125-151.

Pihl, M. (1973). 'The scientific achievements of L.V. Lorenz', *Centaurus*, **17**(1): 83-94.

Pihl, M. (1980). 'Two contributions of L.V. Lorenz to mathematical physics', *Centaurus*, **24**(1): 361-368.

Pipes, L.A. (1940). 'A matrix generalization of Heaviside's expansion theorem', *Journal of the Franklin Institute*, **230**(4): 483-499.

Piquemal, F., B. Jeckelmann, L. Callegaro, J. Hällström, T.J.B.M. Janssen, J. Melcher, G. Rietveld, *et al.* (2017). 'Metrology in electricity and magnetism: EURAMET activities today and tomorrow', *Metrologia*, 54: R1-R24.

Poodiack, R.D. and K.J. LeClair (2009). 'Fundamental theorems of algebra for the perplexes', *The College Mathematics Journal*, **40**(5): 322-335.

Popper, K.R. (1959). *The Logic of Scientific Discovery*. New York: Harper Torchbooks.

Popper, K.R. (1962). *Conjectures and Refutations: The Growth of Scientific Knowledge*. New York: Basic Books.

Popper, K.R. (1992). *Quantum Theory and the Schism in Physics*. London: Routledge.

Post, E.J. (1997). *Formal Structure of Electromagnetics: General Covariance and Electromagnetics*. Mineola, NY: Dover.

Pound, R.V. and G.A. Rebka (1964). 'Apparent weight of electrons', *Physical Review Letters*, **4**(7): 337-341.

Pound, R.V. and J.L. Snider (1964). 'Effect of gravity on nuclear resonance', *Physical Review Letters*, **13**(18): 539-540.

Pound, R.V. and J.L. Snider (1965). 'Effect of gravity on gamma radiation', *Physical Review*, **140**(3B): B788-B803.

Powell, L. (2004). *Power System Load Flow Analysis*. New York: McGraw-Hill.

Poynting, J.H. (1884). 'On the transfer of energy in the electromagnetic field', *Philosophical Transactions of the Royal Society of London*, 175: 343-361.

Poynting, J.H. (1885a). 'On the connexion between electric current and the electric and magnetic inductions in the surrounding field', *Philosophical Transactions of the Royal Society of London*, 176: 277-306.

Poynting, J.H. (1885b). 'On the connexion between electric current and the electric and magnetic inductions in the surrounding field', *Proceedings of the Royal Society of London*, 38: 168-172.

Poynting, J.H. (1909). 'The wave motion of a revolving shaft, and a suggestion as to the angular momentum in a beam of circularly polarised light', *Proceedings of the Royal Society of London, Series A*, 82: 560-567.

Poynting, J.H. (1911). 'On small longitudinal material waves', *Proceedings of the Royal Society of London, Series A*, **85**(580): 474-476.

Preyer, N.W. (2000a). 'Surface charges and fields in simple circuits', *American Journal of Physics*, **68**(11): 1002-1006.

Preyer, N.W. (2000b). 'Transient behavior of simple RC circuits', *American Journal of Physics*, **70**(11): 187-193.

Psarros, N. (1996). 'The mess with the mass terms'. In: *Philosophers in the Laboratory*, (Ed.). V. Mosini, 123-131. Rome: Euroma.

Pugh, E.M. (1961). 'Conservative fields in DC networks', *American Journal of Physics*, **29**(8): 484-486.

Pugh, E.M. and F.E. Pugh (1960). *Principles of Electricity and Magnetism*. London: Addison-Wesley Publ. Co.

Pugh, E.M. and F.E. Pugh (1996). 'Physical significance of the Poynting vector in static fields', *American Journal of Physics*, **35**(2): 153-156.

Punga,   F.   (1901).   'Anwendung   der   Grassmann'schen   linearen

Ausdehnungslehre auf die analytische und graphische Behandlung von Wechselstromerscheinungen', *Zeitschrift für Elektrotechnik*, 42: 506-508; 43: 516-520.

Purcell, E.M. and D.J. Morin (2013). *Electricity and Magnetism*. Cambridge: Cambridge University Press.

Purrington, R.D. (2018). *The Heroic Age: The Creation of Quantum Mechanics, 1925-1940*. Oxford: Oxford University Press.

Puska, P. (2000). 'Note on Gibbs-Heaviside vector 'algebra'', *Internal Report: Electromagnetics Laboratory*, Helsinki University of Technology, Finland. 1-2.

Quade, W. (1934). 'Wirk-, Blind- und Scheinleistung bei Wechselströmen mit beliebiger Kurvenform', *Archiv für Elektrotechnik*, **28**(2): 130-138.

Quade, W. (1939). 'Neue Darstellung der Verzerrungsleistung eines Wechselstroms mit Hilfe des Funktionsraums', *Archiv für Elektrotechnik*, **33**(5): 277-305.

Quade, W. (1940). 'Matrizen rechnung und elektrische Netze', *Archiv für Elektrotechnik*, **34**(10): 545-567.

Quigg, C. (2008). 'The coming revolutions in particle physics', *Scientific American*, **289**(2): 46-53.

Quinn, J.A. (2017). 'A complex quaternion model for hyperbolic 3-space'. ArXiv: 1701.06709 Arxiv.org.

Ramo, S., J.R. Whinnery and T.D. Van Duzer (2013). *Fields and Waves in Communication Electronics*. New Delhi: Wiley.

Rañada, A. F. (1989). 'A topological theory of the electromagnetic field', *Letters in Mathematical Physics*, **18**(2): 97-106.

Rañada, A.F. (1990). 'Knotted solutions of the Maxwell equations in vacuum', *Journal of Physics A: Mathematical and General*, **23**(16): 815-820.

Rañada, A.F. (1992). 'Topological electromagnetism', *Journal of Physics A: Mathematical and General*, **25**(6): 1621-1641.

Rañada, A.F. (2012). 'On topology and electromagnetism', *Annalen der Physik*, 524(2): 35-37.

Rañada, A.F. and J.L. Trueba (2001). 'Topological electromagnetism with hidden nonlinearity'. In: *Modern Nonlinear Optics Part 3*, 2nd ed., (Ed.). M.W. Evans, 197-254, *Advances in Chemical Physics*, 119. John Wiley & Sons, Inc.

Rangacharyulu, C. (2015). 'An epitaph for all photons: A phoenix rising from its ashes', *Proceedings of the Society of Photo-Optical Instrumentation Engineers (SPIE) 9570, The Nature of Light: What are Photons?* VI, 957001.

Rashkovskiy, S.A. (2015). 'Are there photons in fact?' *Proceedings of the Society of Photo-Optical Engineers (SPIE) 9570, The Nature of Light: What are Photons?* VI, 95700G. 1-13.

Rechenberg, H. (1997). 'The electron in physics: Selection from a chronology of the last 100 years', *European Journal of Physics*, **18**(3): 145-149.

Redhead, M. (1995). *From Physics to Metaphysics. The Tarner Lectures,* Cambridge Trinity College, February 1993. Cambridge: Cambridge University Press.

Redlich, R. (1984). 'Note on power and Poynting vector in low-frequency circuits', *IEEE Transactions on Education,* **E-27**(2): 109.

Reibiger, A. (2011). 'Foundations of network theory', *COMPEL, The International Journal for Computation and Mathematics in Electrical and Electronic Engineering,* **30**(4): 1319-1332.

Reif, F. (1982). 'Generalized Ohm's law, potential difference, and voltage measurements', *American Journal of Physics,* **50**(11): 1048-1049.

Renton, P. (1990). *Electroweak Interactions: An Introduction to the Physics of Quarks and Leptons.* Cambridge: Cambridge University Press.

Resendes, D.P. (2013). 'Geometric algebra in plasma electrodynamics', *Journal of Plasma Physics,* **79**(5): 735-738.

Riemann, B. (1854). 'On the hypotheses which lie at the bases of geometry', translated by W.K. Clifford, *Nature,* **8**(183): 14-17; (184): 36-37.

Riesz, M. (1993). *Clifford Numbers and Spinors* (Ed. E.F. Bolinder and P. Lounesto. facsimile reprint of Riesz's 1957-1958 lectures at the University of Maryland. Dordrecht: Kluwer Academic Publishers.

Robinson, F.N.H. (1975). 'Electromagnetic stress and momentum in matter', *Physics Reports* (Section C of Physics Letters), **16**(6): 313-354.

Robinson, V.N.E. (2011). 'A proposal for the structure and properties of the electron', *Particle Physics Insights,* 4: 1-18. https://doi.org/10.4137/PPI.S7102.

Rocha, V.D., C. Zagoya and M.M. Mares (2008). 'Poynting's theorem for plane waves at an interface: A scattering matrix approach', *American Journal of Physics,* **76**(7): 621-625.

Roche, J. (1998). 'The present status of Maxwell's displacement current', *European Journal of Physics,* **19**(2): 155-166.

Roche, J. (2003). 'What is potential energy?' *European Journal of Physics,* **24**(2): 185-196.

Roche, J. (2006). 'What is momentum?' *European Journal of Physics,* **27**(5): 1019-1036.

Rodrigues, O. (1840). 'Des lois géométriques qui régissent les déplacements d'un système solide dans l'espace, et de la variation des coordonnées provenant de ces déplacements considérés indépendamment des causes qui peuvent les produire', *Journal de Mathématiques Pures et Appliquées,* 5: 380-440.

Rodrigues, W.A. and E.C. de Oliveira (2007). *The Many Faces of Maxwell, Dirac and Einstein Equations: A Clifford Bundle Approach,* The Lecture Notes in Physics, 722. Berlin: Springer.

Rohrlich, F. (1973). 'The electron: Development of the first elementary particle theory'. In: *The Physicist's Conception of Nature,* (Ed.). J. Mehra, 331-369. Dordrecht: D. Reidel Publishing Co.

Rohrlich, F. (1974). 'The nonlocal nature of electromagnetic interactions'. In: *Physical Reality and Mathematical Description*, (Ed.). J. Mehra, 387-402. Dordrecht: D. Reidel Publishing Co.

Rohrlich, F. (1988). 'Pluralistic ontology and theory reduction in the physical sciences', *The British Journal for the Philosophy of Science*, **39**(3): 295-312.

Rohrlich, F. (2007). *Classical Charged Particles*, 3rd ed. Hackensack, NJ: World Scientific.

Romer, R.H. (1982a). 'Alternatives to the Poynting vector for describing the flow of electromagnetic energy', *American Journal of Physics*, **50**(12): 1166-1168.

Romer, R.H. (1982b). 'What do "voltmeters" measure? Faraday's law in a multiply connected region', *American Journal of Physics*, **50**(12): 1089-1093.

Rooney, J. (1978). 'On three types of complex number and planar transformations', *Environment & Planning B*, 5: 89-99.

Rooney, J. (2014). 'Generalised complex numbers in mechanics'. In: *Advances on Theory and Practice of Robots and Manipulators*, (Ed.). M. Ceccarelli and V.A. Glazunov, 55-62. Springer International Publishing, Switzerland.

Rosser, W.G.V. (1963). 'What makes an electric current 'flow'?', *American Journal of Physics*, **31**(11): 884-885.

Rosser, W.G.V. (1968). *An Alternative Approach to Maxwell's Equations: Classical Electromagnetism via Relativity*. Springer Science+Business Media.

Rosser, W.G.V. (1970). 'Magnitudes of surface charge distributions associated with electric current flow', *American Journal of Physics*, **38**(2): 265-266.

Rosser, W.G.V. (1976). 'Does the displacement current in empty space produce a magnetic field?' *American Journal of Physics*, **44**(12): 1221-1223.

Rota, G-C. [1997] (2008). *Indiscrete Thoughts*. Boston: Birkhäuser.

Rotella, F. and I. Zambettakis (2006). 'Du calcul opérationnel à l'opérateur de transfert', *e-STA - Revue des Sciences et Technologies de l'Automatique*, **3**(4): 1-9.

Rotella, F. and I. Zambettakis (2013). 'An operational standpoint in electrical engineering', *Electronics*, **17**(2): 71-81.

Roth, S. (2007). *Precision Electroweak Physics at Electron-Positron Colliders*. Berlin: Springer.

Rothe, F.S. (1953). *An Introduction to Power System Analysis*. New York: Wiley.

Rousseaux, G., R. Kofman and O. Minazzoli (2008). 'The Maxwell-Lodge effect: Significance of electromagnetic potentials in the classical theory', *The European Physical Journal D - Atomic, Molecular, Optical and Plasma Physics*, 49: 249-256.

Rumpel, D. and Sun J.R. (1989). *Netzleittechnik*. Berlin: Springer.

Russell, B. (1929). *Marriage and Morals*. New York: Liveright.

Russell, B.R. (1968). 'Surface charges on conductors carrying steady currents', *American Journal of Physics*, **36** (6): 527-529.

Russell, J.B. (1942). 'Heaviside's direct operational calculus', *Electrical Engineering*, **61**(2): 84-88.

Russer, P. (2003). *Electromagnetics, Microwave Circuit and Antenna Design for Communications Engineering*. Boston: Artech House.

Rutherford, E. (1911). 'LXXIX. The scattering of α and β particles by matter and the structure of the atom', *The London, Edinburgh, and Dublin Philosophical Magazine and Journal of Science* Series 6, **21**(125): 669-688.

Saa, A. (2011). 'Local electromagnetic duality and gauge invariance', *Classical and Quantum Gravity*, **28**(12): 127002.

Saá, D. (2007). 'Four Vector Algebra'. arXiv:0711.3220v1 [math-ph] 20 Nov. 2007.

Saadat, H. (1999). *Power System Analysis*. Boston: McGraw-Hill.

Sabbata, V. and B.K. Datta (2007). *Geometric Algebra and Applications to Physics*. New York: Taylor & Francis.

Sah, A. P-T. (1936a). 'Dyadic algebra applied to 3 phase circuits', *Transactions of the American Institute of Electrical Engineers*, **55**(8): 876-882.

Sah, A. P-T. (1936b). 'Complex vectors in 3-phase circuits', *Transactions of the American Institute of Electrical Engineers*, **55**(12): 1356-1364.

Sah, A. P-T. (1936c). 'Discussions of AIEE Papers – as Recommended for Publication by Technical Committee', [Discussion of two above papers by Sah, P-T]. *Transactions of the American Institute of Electrical Engineers*, **55**(12): 610-612.

Sah, A. P-T. (1936d). 'Author's closure', *Transactions of the American Institute of Electrical Engineers*, **55**(8): 1030-1031.

Sah, A. P-T. (1939). *Dyadic Circuit Analysis*. Scranton, Pa.: International Textbook Company.

Salam, Abdus (1980). "Gauge unification of fundamental forces', *Science*, **210**(4471): 723-732.

Salmeron, P. and R.S. Herrera (2009). 'Instantaneous reactive power theory: A general approach to poly-phase systems', *Electric Power Systems Research*, 79: 1263-1270.

Sánchez-Ron, J.M. (2009). 'Memories of old times: Schlick and Reichenbach on time in quantum mechanics', *Lecture Notes in Physics*, 789: 1-13.

Sangston, K.J. (2016). 'Geometry of complex data', *IEEE Aerospace and Electronic Systems Magazine*, **31**(3): 32-69.

Sangwine, J., T.A. Ell and N. Le Bihan (2011). 'Fundamental representations and algebraic properties of biquaternions or complexified quaternions', *Advances in Applied Clifford Algebras*, **21**(3): 607-636.

Sauer, P.W. (2005). 'Reactive power and voltage control issues in electric power systems'. In: *Applied Mathematics for Restructured Electric Power Systems* (Ed.). J.H. Chow, F.F. Wu and J.A. Momoh, 11-24. Springer.

Sauer, P.W. and M.A. Pai (1998). *Power System Dynamics and Stability*. Upper Saddle River, NJ: Prentice Hall.

Schaller, D. (1972). *Maschinelle Berechnung und Optimierung*. Leipzig: *VEB Deutscher Verlag für Grundstoffindustrie*.

Schelkunoff, S.A. (1948). 'Methods of electromagnetic field analysis', *The Bell System Technical Journal*, **27**(3): 487-509.

Schelkunoff, S.A. (1955). 'Conversion of Maxwell's equations into generalized telegraphist's equations', presented at a meeting of International Scientific Radio Union on May 4, 1954 in Washington, D.C., *The Bell System Technical Journal*, **34**(55): 995-1043.

Scholz, E. (Ed.) (1990). *Geschichte der Algebra*. Mannheim: *Bibliographisches Institut* & F.A. Brockhaus AG,.

Schönfeld, E. (1990). 'Electron and fine-structure constant', *Metrologia*, 27: 117-125.

Schönfeld, E. and P. Wilde (2008). 'Electron and fine-structure constant II', *Metrologia* 45: 342-355.

Schönfeld, H. (1951). *Die wissenschaftlichen Grundlagen der Elektrotechnik*. Leipzig: S. Hirzel Verlag.

Schultheiß, F. and K.-D. Weßnigk (1971). *Berechnung elektrischer Energieversorgungsnetze*, vol. 2: *Übertragungsberechnung*. Leipzig: *VEB Deutscher Verlag für Grundstoffindustrie*.

Schwab, A.J. (2012). *Elektroenergiesysteme. Erzeugung, Transport, Übertragung und Verteilung elektrischer Energie*. Heidelberg: Springer.

Schwinger, J., L.L. DeRaad, K.A. Milton and Wu-yang Tsai (1998). *Classical Electrodynamics*. Reading (UK): Perseus Books.

Scott, R.E. (1960). *Linear Circuits*. Reading, MA: Addison-Wesley Publishing Co.

Seely, S. (1984). 'Poynting's theorem and energy-flow postulate', *IEEE Transactions on Education*, **E-27**(4): 246.

Selig, J.M. (2005). *Geometric Fundamentals of Robotics*. New York: Springer.

Sellin, I. (1982). 'Atomic structure and spectra', *McGraw-Hill Encyclopedia of Science and Technology*, vol. I: 857.

Semon, M.D. and J.R. Taylor (1996). 'Thoughts on the magnetic vector potential', *American Journal of Physics*, **64**(11): 1361-1369.

Sen, D. and D. Sen (2016). 'Representation of physical quantities: From scalars, vectors, tensors and spinors to multivectors'. https://www.researchgate.net/publication/310604392.

Serrano-Iribarnegaray, L. (1993). 'The modern space-phasor theory, Part I: Its coherent formulation and its advantages for transient analysis of converter-fed AC machines', *European Transactions on Electrical Power*, 3(2): 171-180.

Serrano-Iribarnegaray, L. (2000). 'The space phasor theory'. In: *Modern Electrical Drives*, (Ed.). H.B. Ertan, M. Y. Üctug, R. Colyer and A. Consoli, 393-423. NATO, ASI Series. Kluwer Academic Publishers.

Shchedrin, G. (2013). 'The Schrödinger description of a single proton', *Proceedings of the Society of Photo-Optical Instrumentation Engineers (SPIE), The Nature of Light: What are Photons?* V, 883213 (1 October 2013). 6 p.

Shepherd, W. and P. Zakikhani (1972). 'Suggested definition of reactive power for non-sinusoidal systems', *Proceedings of the Institution of Electrical Engineers*, 119(9): 1361-1362.

Shepherd, W. and P. Zakikhani (1973). 'Suggested definition of reactive power for non-sinusoidal systems. Discussion. Answer to E. Micu', *Proceedings of the Institution of Electrical Engineers*, 120(7): 797-798.

Sherwood, B.A. and R.W. Chabay (2009). 'A unified treatment of electrostatics and circuits'.https://matterandinteractions.org/wp-content/uploads/2016/07/circuit.pdf.

Silagadze, Z. (2002). 'Multidimensional vector product', *Journal of Physics A, Mathematics and General*, 35(23): 4949-4953.

Silberstein, L. (1907a). 'Elektromagnetische Grundgleichungen in bivectorieller Behandlung', *Annalen der Physik*, 327(3): 579-586.

Silberstein, L. (1907b). 'Nachtrag zur Abhandlung über Elektromagnetische Grundgleichungen in bivektorieller Behandlung', *Annalen der Physik*, 329(14): 783-784.

Silberstein, L. (1914). *The Theory of Relativity*. London: Macmillan.

Silberstein, L. (1922). *The Theory of General Relativity and Gravitation*. Toronto: University of Toronto Press.

Silsbee, F.B. (1920). 'Power factor in polyphase systems', *Journal of the American Institute of Electrical Engineers*, 39(6): 542-543.

Simons, P. (2009). 'Vectors and beyond: Geometric algebra and its philosophical significance', *Dialectica*, 63(4): 381-395.

Simony, K. (1956). *Theoretische Elektrotechnik*. Berlin: *VEB, Deutscher Verlag der Wissenschaften*.

Simony, K. (2012). *A Cultural History of Physics*. Boca Raton, FL: CRC Press.

Slepian, J. (1919). 'Inherent limitations on transformations possible by stationary apparatus', *Transactions of the American Institute of Electrical Engineers*, 38(2): 1697-1711.

Slepian, J. (1920). 'Reactive power and magnetic energy', presented at the 36th Annual Convention of the American Institute of Electrical Engineers, White Sulphur Springs, June 30: 1115-1132.

Slepian, J. (1942). 'Energy flow in electric systems: The *Vi* energy-flow postulate', *Transactions of the American Institute of Electrical Engineers*, 61(12): 835-841.

Slepian, J. (1942). 'Energy and energy flow in the electromagnetic field', *Journal of Applied Physics*, 13: 512-518.

Slepian, J. (1950). 'Electromagnetic ponderomotive forces within material bodies', *Proceedings of the National Academy of Sciences of the United States of America*, **36**(9): 485-497.

Slonim, M.A. (2004). 'Physical essence of power components', *L'Energia Elettrica*, 81: 91-96.

Slonim, M.A. and Wyk Van, J.D. (1988). 'Power components in a system with sinusoidal and non-sinusoidal voltages and/or currents', *IEE Proceedings B - Electric Power Applications*, **135**(2): 76-84.

Smith, C.F. (1934). *Practical Alternating Currents and Alternating Current Testing*. Manchester: The Scientific Publishing Co.

Smith, J.J. and P.L. Alger (1950). 'Justification of Heaviside methods', *Electrical Engineering*, **69**(2): 116-116.

Sobczyk, G. (1985). 'The hyperbolic number plane', *The College Mathematics Journal*, **26**(4): 268-280.

Sobczyk, G. (1996). 'Introduction to Geometric Algebra'. In: *Clifford (Geometric) Algebras*, (Ed.). W.E. Baylis, 37-43. Boston: Birkhäuser.

Sobczyk, G. (2008). 'Geometric matrix algebra', *Linear Algebra and its Applications*, **429**(5 + 6): 1163-1173.

Sobczyk, G. (2013a). 'Special relativity in complex vector algebra'. arXiv:0710.0084v1 [math-ph] 29 Sep. 2007.

Sobczyk, G. (2013b). *New Foundations in Mathematics: The Geometric Concept of Number*. New York: Birkhäuser.

Sobczyk, G. (2015). 'Geometry of spin ½ particles', *Revista Mexicana de Fisica*, **61**(3): 211-223.

Sobczyk, G. (2019). *Matrix Gateway to Geometric Algebra, Space-time and Spinors*. San Bernardino, CA: Independently published [ISBN 978 1704 596624].

Solymar, L. (1984). *Lectures on Electromagnetic Theory*,.Oxford Science Publications. Oxford University Press.

Someda, C.G. (2006). *Electromagnetic Waves*. Boca Raton, FL: CRC Press.

Sommerfeld, A. (1928). 'Zur Elektronentheorie der Metalle', *Naturwissenschaften*, 16: 374-381.

Späth, H. (2007). 'A general purpose definition of active current and non-active power based on German standard DIN 40110', *Electrical Engineering*, 89: 167-175.

Spring, E. (2003). *Elektrische Energienetze*. Berlin: VDE Verlag GMBH.

Stagg, G.W. and El-Abiad, A.H. (1968). *Computer Methods in Power System Analysis*. Singapore: McGraw-Hill Book Co.

Stahlhut, J.W., T.J. Browne and G.T. Heydt (2007). 'The assessment of the measurement of the Poynting vector for power system instrumentation', 2007, 39th North American Power Symposium, Las Cruces, NM, 30 September-2 October. 237-242.

Stankovic, A.M. and H. Lev-Ari (2000). 'Frequency-domain observations on definition of reactive power', *IEEE Power Engineering Review*, **20**(6): 46-48.

Stankovic, A.M., S.R. Sanders and T. Aydin (2002). 'Dynamic phasors in modeling and analysis of unbalanced polyphase AC machines', *IEEE Transactions on Energy Conversion*, **17**(1): 107-113.

Staudt, V. (2002). 'Power Theory: Power currents, active currents, non-active currents' [lecture notes]. Bochum, Germany: Institute for Generation and Application of Electrical Energy. 1-25.

Staudt, V. (2008). 'Fryze-Buchholz-Depenbrock: A time-domain power theory', *International School on Nonsinusoidal Currents and Compensation*, Lagow, Poland: 1-12.

Staudt, V. and H. Wrede. 'On the compensation of non-active current components of three-phase loads with quickly changing unsymmetry', *European Transactions on Electrical Power*, **11**(5): 301-307.

Steck, B. (2008). *Application en métrologie électrique de dispositifs monoélectroniques: Vers une fermeture du triangle métrologique*, doctoral thesis,Université de Caen. https://tel.archives-ouvertes.ft/tel-00203161.

Steger, U., U. Büdenbender, E. Feess and D. Nelles (2008). *Die Regulierung elektrischer Netze*. Berlin: Springer.

Steinmetz, Ch. (1890). 'Transformatoren-problem in elementar-geometrischer Behandlungsweise', *ETZ Elektrotechnische Zeitschrift*, 11: 185-186; 205-206; 225-227, 233-234; 345.

Steinmetz, Chas (1891a). 'Anwendung des Polardiagrams der Wechselströme für induktive Widerstände', *ETZ Elektrotechnische Zeitschrift*, 12: 394-396.

Steinmetz, C.P. (1891b). 'Elementary geometrical theory of the alternate-current transformer', *Electrical Engineer*, 11: 627-629.

Steinmetz, C.P. (1892a). 'On the curves which are self-reciprocal in a linear nulsystem, and their configurations in space', *American Journal of Mathematics*, **14**(2): 161-186.

Steinmetz, C.P. (1892b). 'The energy-function of the magnetic circuit', *Science*, **20**(509): 258-259.

Steinmetz, C.P. (1893). 'Complex quantities and their use in electrical engineering', *Proceedings of the International Electrical Congress*, Chicago, August 21st to 25th, 33-74.

Steinmetz, C.P. (1897). *Theory and Calculation of Alternating Current Phenomena* (with the assistance of E.J. Berg). New York: The W.J. Johnston Company.

Steinmetz, C.P. (1899). 'Symbolic representation of general alternating waves and of double frequency vector products', *Transactions of the American Institute of Electrical Engineers*, 16: 269-296.

Steinmetz, C.P. (1900a). *'Theorie und Berechnung der Weschselstromerscheinungen'*. Berlin: Verlag von Reuther & Reichard.

Steinmetz, C.P. (1900b). *Theory and Calculation of Alternating Current Phenomena* (with the assistance of Ernst Berg), 3rd ed. New York: Electrical World and Engineer Inc.

Steinmetz, C.P. (1909). Discussion on 'Output and regulation in long-distance lines', *American Institute of Electrical Engineers*, June 29: 691-693.

Steinmetz, C.P. (1911). 'Some unexplored fields in electrical engineering', *Journal of the Franklin Institute of the State of Pennsylvania*, **171**(6): 537-559.

Steinmetz, C.P. (1912). *Théorie et Calcul des Phénomènes Électriques de Transitions et des Oscillations*. Paris: *H. Dunot et E. Pinat*.

Stephenson, R.J. (1966). 'Development of vector analysis from quaternions', *American Journal of Physics*, **34**(3): 194-201.

Stepina, J. (1989). 'Komplexe Größen in der Elektrotechnik', *Archiv für Elektrotechnik*, **72**(6): 407-414.

Stevenson, W.D. (1982). *Elements of Power Analysis*, 4th ed. New York: McGraw-Hill.

Stewart, A.M. (2005a). 'Angular momentum of light', *Journal of Modern Optics*, **52**(8): 1145-1154.

Stewart, A.M. (2005b). 'Angular momentum of the electromagnetic field: The plane wave paradox resolved', *European Journal of Physics*, **26**(4): 635-641.

Stock, M. (2011). 'The watt balance: Determination of the Planck constant and redefinition of the kilogram', *Philosophical Transactions of the Royal Society A: Mathematical, Physical and Engineering Sciences*, **369**(1953): 3936-3953.

Stocklmayer, S. and D.F. Treagust (1994). 'A historical analysis of electric currents in textbooks: A century of influence on physics education', *Science & Education*, **3**(2): 131-154.

Stockman, H. (1956). *The jw– or Symbolic Method: A Guide for the Practical Engineer in the Use of the Symbolic Method with Applications in the Fields of Radio and Electronics*, SERCO Technical Series 101. Waltham, MA: SER Co.

Stone, M. (2000). 'Phonons and forces: Momentum versus pseudo-momentum in moving fluids'. arXiv: cond-mat/0012316v1 17 Dec.: 1-30.

Stratton, J.A. [1941] (2007). *Electromagnetic Theory*. Piscataway, NJ: The IEEE Press.

Sugon, Q.M. and D.J. McNamara (2008). 'Electromagnetic energy-momentum equations without tensors: A geometric algebra approach'. arXiv:0807.1382v1 [physics.class-ph] 9 Jul.: 1-4.

Sumpner, W.E. (1932). 'Electromagnetic waves and pulses', *The London, Edinburgh, and Dublin Philosophical Magazine and Journal of Science,* Series 7, **13**(87): 1049-1075.

Suter, J. (2003). 'Geometric algebra primer'. http://www.jaapsuter.com/paper/ga_primer.pdf.

Sutherland, P.E. (2007). 'On the definition of power in an electrical system', *IEEE Transactions on Power Delivery,* **22**(2): 1100-1107.

Sutton, A.M. (1967). 'Spinning particles', *Physical Review,* **160**(5): 1055-1064.

Swartz, Ch. W. (1965). *An Operator Calculus Based on the Laplace Transform,.* Ph.D. thesis (65-9922), University of Arizona.

Tai, C-T. (1997). *Generalized Vector and dyadic Analysis: Applied Mathematics in Field Theory.* Piscataway, NJ: IEEE Press.

Tait, P.G. (1886). 'Quaternions', *Encyclopedia Britannica,* 9th ed. , vol. 20: 160-164.

Tanabashi, M. *et al.* (Particle Data Group) (2018). 'Review of Particle Physics', *Physical Review,* D 98:030001. 1-1898.

Tenti, P. and P. Mattavelli (2004). 'A time-domain approach to power term definitions under non-sinusoidal conditions', *L'Energia Elettrica,* 81: 75-84.

Tevian, D. (2012). *The Geometry of Special Relativity.* Boca Raton, FL: CRC Press.

Thidé, B. (2013). *Electromagnetic Field Theory,* 2nd ed. ISBN 978-0-486-4773-2.

Thomson, J.J. (1897). 'Cathode rays', *Proceedings of the Royal Institution [of Great Britain],* 15: 419-432.

Thompson, S.P. (1895). *Polyphase Electric Currents.* New York: E & F.N. Spon.

Timar, R.L., I. Schmidt and G. Retter (2000). 'Space vector theory'. In: *Modern Electrical Drives,* (Eds.) H.B. Ertan, M.Y. Üctug, R. Colyer and A. Consoli, 359-392, NATO, ASI Series. Kluwer Academic Publishers.

Todeschini, G., A.E. Emanuel, A. Ferrero and A.P. Morando (2007). 'A Poynting vector approach to the study of the Steinmetz compensator', *IEEE Transactions on Power Delivery,* **22**(3): 1830-1833.

Tong, D. (2015). *Lectures on Electromagnetism.* University of Cambridge. http://www.damtp.cam.ac.uk/user/tong/em.html.

Tran, M. (2018). 'Evidence for Maxwell's equations, fields, force laws and alternative theories of classical electrodynamics', *European Journal of Physics,* **39**(6): 063001.

Trueba, J.L. and A.F. Rañada (1996). 'The electromagnetic helicity', *European Journal of Physics,* **17**(3): 141-144.

Ţugulea, A. (1996). 'Criteria for the definition of the electric power quality and its measurement systems', *European Transactions on Electrical Power,* **6**(5): 357-363.

Țugulea, A. (1998). 'Power flows in distorted electromagnetic fields', *Proceedings of the 8th International Conference on Harmonics and Quality of Power, Athens, Greece* (Cat. No.98EX227), vol. 2, 678-684. https://doi.org/10.1109/ICHQP.1998.760127.

Țugulea, A. (2002). 'Some remarks concerning the measurement of the ferromagnetic losses under non-sinusoidal conditions', *European Transactions on Electrical Power*, **12**(2): 85-92.

Tung, W-K. (1985). *Group Theory in Physics*. Singapore: World Scientific Publishing Co.

Ungar, A.A. (2008). *Analytic Hyperbolic Geometry and Albert Einstein's Special Theory of Relativity*. Hong Kong: World Scientific.

Ungar, A.A. (2014). *Analytic Hyperbolic Geometry: Mathematical Foundations and Applications*. Hong Kong: World Scientific.

Valkenburg, M.E. (1964). *Network Analysis*. Englewood Cliffs, NJ: Prentice Hall.

Vallarta, M.S. (1926). 'Heaviside's proof of his expansion theorem', *Journal of the American Institute of Electrical Engineers*, **45**(4): 383-387.

Van Belle, J.L. (2019). 'The emperor has no clothes: A realist interpretation of quantum mechanics'. https://vixra.org/abs/1901.0105.

Van Bladel, J.G. (2007). *Electromagnetic Fields*. Piscataway, NJ: IEEE Press.

Vandenbosch, Guy A.E. (2013). 'How to model connection wires in a circuit: From physical vector fields to circuit scalar quantities', *American Journal of Physics*, **81**(9): 676-681.

Vanderlinde, J. (2004). *Classical Electromagnetic Theory*. New York: Kluwer Academic Publishers.

van der Mark, M.B. (2015a). 'On the nature of 'stuff' and the hierarchy of forces', *Proceedings of the Society of Photo-Optical Instrumentation Engineers (SPIE) 9570, The Nature of Light: What are Photons?* VI, 95701G. 15 p.

van der Mark, M.B. (2015b). 'Quantum mechanical probability current as electromagnetic 4-current from topological EM fields', *Proceedings of the Society of Photo-Optical Instrumentation Engineers (SPIE) 9570, The Nature of Light: What are Photons?* V, 95701I-16. spiedigitallibrary.org.

van der Mark, M.B. (2015c). 'Topological electromagnetism is the stuff that matters!' *Proceedings of the Society of Photo-Optical Instrumentation Engineers (SPIE) 9570, The Nature of Light: What are Photons?* V, 95701I-16.

van der Mark, M.B. and G.W. 't Hooft (2000). 'Light is heavy'. arXiv:1508.06478 [physics.hist-p].

van der Pol, Balthasar (1929). 'On the operational solution of linear differential equations and an investigation of the properties of these solutions', *The London, Edinburgh, and Dublin Philosophical Magazine and Journal of Science*, Series 7. 8 (53 (Supplement): 861-898.

van der Pol, Balthasar and K.F. Niessen (1932). 'Symbolic calculus', *The London, Edinburgh, and Dublin Philosophical Magazine and Journal of Science*, Series 7, **13**(85): 537-577.

van Fraassen, B.C. (1980). *The Scientific Image*. Oxford: Clarendon Press.

Varney, R.N. and L.H. Fisher (1984). 'Electrical fields associated with stationary currents', *American Journal of Physics*, **52**(12): 1097-1099.

Vas, P. and J.L. Willems (1987). 'The application of space-vector theory to the analysis of electrical machines with space harmonics', *Archiv für Elektrotechnik*, **70**(5): 359-363.

Vaschy, A. (1890). *Traité d'Électricité et de Magnétisme*. Paris: Baudry.

Vaz, J., Jr. and R. da Rocha (2016). *An Introduction to Clifford Algebras and Spinors*. Oxford: Oxford University Press.

Veblen, O. (1934). 'Spinors', *Science*, **80**(2080): 415-419.

Vickers, P. (2008). 'Frisch, Muller and Belot on an inconsistency in classical electrodynamics', *The British Journal for the Philosophy of Science*, **59**(4): 767-792.

Vickers, P. (2013). *Understanding Inconsistent Science*. Oxford University Press.

Vickers, P. (2014). 'Theory flexibility and inconsistency in science', *Synthese*, **191**(13): 2891-2906.

Vince, J. (2008). *Geometric Algebra for Computer Graphics*. London: Springer.

Vince, J. (2011). *Quaternions for Computer Graphics*. London: Springer.

Vince, J. (2017). *Mathematics for Computer Graphics*. London: Springer.

Vistnes, A.I. (2013). 'The problematic photon', *Proceedings of the Society of Photo-Optical Instrumentation Engineers (SPIE) 8832, The Nature of Light: What are Photons?* V, 883213 (1 October 2013). https://doi.org/10.1117/12.2023881. 9 p.

Vold, T.G. (1993a). 'An introduction to geometric algebra with an application in rigid body mechanics', *American Journal of Physics*, **61**(6): 491-504.

Vold, T.G. (1993b). 'An introduction to geometric calculus and its application to electrodynamics', *American Journal of Physics*, **61**(6): 505-513.

Wadhwa, C.L. (1992). *Electrical Power Systems*. New Delhi: Wiley Eastern Limited.

Waerden, van der, B.L. (1976). 'Hamilton's discovery of Quaternions', *Mathematics Magazine*, **49**(5): 227-234.

Wagner, C.F. and R.D. Evans (1933). *Symmetrical Components*. New York: McGraw-Hill Book Co., Inc.

Wagner, K.W. (1916). 'Über eine Formel von Heaviside zur Berechnung von Einschaltvorgängen (Mit Anwendungsbeispielen)', *Archiv für Elektrotechnik*, **4**(5): 159-193.

Wagner, K.W. (1941). 'Laplacesche Transformation und Operatorenrechnung', *Archiv für Elektrotechnik*, **35**(8): 502-506.

Warnick, K.F. and P. Russer (2006). 'Two, three and four-dimensional electromagnetics using differential forms', *Turkish Journal of Electrical Engineering and Computer Science*, **14**(1): 153-171.

Watson, D. (1936). 'Discovery of positron one of science's great events', *The Science Newsletter*, **30**(815): 324-325.

Weber, W. (1846). *Elektrodynamische Maassbestimmungen*. Leipzig: *Weidmann'sche Buchhandlung*.

Weber, W. (1872). 'Electrodynamic measurements. Sixth Memoir, relating specially to the Principle of the Conservation of Energy', *The London, Edinburgh, and Dublin Philosophical Magazine and Journal of Science* (4th ser.) **43**(283): 1-20; (284): 119-149.

Weedy, B.M. and B.J. Cory (1998). *Electric Power Systems*. Chichester: Wiley.

Weinberg, Steven (1977). 'The forces of Nature: Gauge field theories offer the prospect of a unified view of the four kinds of natural force – the gravitational and electromagnetic, and the weak and the strong', *American Scientist*, **65**(2): 171-176.

Weinberg, S. (1976). 'The forces of nature', *Bulletin of the American Academy of Arts and Sciences*, **29**(4): 13-29.

Weinberg, Steven (1980). 'Conceptual Foundations of the Unified Theory of Weak and Electromagnetic Interactions', *Science*, **210**(4475): 1212-1218.

Weiss, P. (2004). 'The electron's other charge: Workhorse of electricity shows its weak side', *Science News*, 165: 278.

Weisstein, E.W. (1999). *CRC Concise Encyclopedia of Mathematics*. Boca Raton, FL: CRC Press.

Wessel, C. [1797] (1999). *On the Analytical Representation of Direction*, (Eds.). B. Branner and J. Lützen. Copenhagen: C.A. ReitzelsForlag.

Westinghouse Electric Corporation (1944). *Electrical Transmission and Distribution Reference Book*, 3rd ed. East Pittsburgh, PA: Westinghouse Electric & Manufacturing Company.

Whitehead, J.B. (1916). Discussion on 'Application of a polar form of complex quantities to the calculation of A-C. phenomena' (N.S. Diamant), Cleveland, Ohio, June 30, 1916 (*see Proceedings* for June, 1916), *Proceedings of the American Institute of Electrical Engineers*, **35**(10): 1529-1531. https://doi.org/10.1109/PAIEE.1916.6590719.

Whittaker, E.T. (1910). *A History of the Theories of Aether and Electricity*, vol 1: *The Classical Theories*. New York: Harper & Brothers.

Whittaker, E.T. (1929). 'What is Energy?' *The Mathematical Gazette*, **14**(200): 401-406.

Widder, D.V. (1945). 'What is the Laplace transform?' *The American Mathematical Monthly*, **52**(8): 419-425.

Wiederkehr, K.H. (2010). 'Über Vorstellungen von der elektrischen

Leitung, die Entwicklung einer Elektronentheorie der Metalle und der Begin einer Festkörperphysik', *Sudhoffs Archiv*, **94**(1): 57-72.

Wiener, N. (1926). 'The operational calculus', *Mathematische Annalen*, 95: 557-584. https://doi.org/10.1007/BF01206627.

Wiener, N. (1933). *The Fourier Integral and Certain of Its Applications*. New York: Dover Publications, Inc.

Wiener, N. (1976). *Norbert Wiener. Collected Works with Commentaries*, vol. 1, (Ed.). P. Masani. Cambridge, MA: MIT Press.

Wiener, N. (1979a). *Norbert Wiener. Collected Works with Commentaries*, vol. 2, (Ed.). P. Masani. Cambridge, MA: MIT Press.

Wiener, N. (1979b). 'The Operational Calculus'. In: *Norbert Wiener. Collected Works with Commentaries*, vol. 2, (Ed.). P. Masani 397-427. Cambridge, MA: MIT Press.

Wiener, N. (1994). *The Legacy of Norbert Wiener: A Centennial Symposium in Honor of the 100th Anniversary of Norbert Wiener's Birth*, (Ed.) D. Jerison, I.M. Singer and D.W. Strook. Cambridge, MA: MIT Press.

Wigner, E.P. (1960). 'The unreasonable effectiveness of mathematics in the natural sciences', *Communications on Pure and Applied Mathematics*, **13**(1): 1-14.

Wilczek, F. (2005). 'In search of symmetry lost', *Nature*, **433**(7023): 239-247.

Wilczek, F. (2013). 'The enigmatic electron', *Nature*, **498**(7452): 31-32.

Willems, J.L. (1996). 'Mathematical foundations of the instantaneous power concepts: A geometrical approach', *European Transactions on Electrical Power*, **6**(5): 299-304.

Willems, J.L. (2004). 'Reflections on apparent power and power factor in non-sinusoidal and polyphase situations', *IEEE Transactions on Power Delivery*, **19**(2): 835-840.

Willems, J.L. (2008). 'Critical analysis of different current decomposition and compensation schemes', *International School on Non-sinusoidal Currents and Compensation*, Lagow, Poland, 2008: 1-6, IEEE 978-1-4244-2130-5.

Willems, J.L. (2010). 'The fundamental concepts of power theories for single-phase and three-phase voltages and currents', *IEEE Instrumentation & Measurement Magazine*, **13**(5): 37-44.

Willems, J.L. (2010). 'Reflections on power theories for poly-phase non-sinusoidal voltages and currents', *International School on Non-sinusoidal Currents and Compensation*, Lagow, Poland: 1-6. IEEE 978-4244-1-5435-8: 5-16.

Willems, J. and J. Ghijselen (2004). 'The Relation Between the Generalized Apparent Power and the Voltage Reference', *L'Energia Elettrica*, **81**(5-6) (research supplement): 37-45.

Williamson, J.G. (2015). 'The nature of the photon and the electron', *Proceedings of the Society of Photo-Optical Instrumentation Engineers*

(*SPIE*) *9570, The Nature of Light: What are Photons?* VI, 957015 (10 September 2015).

Williamson, J.G. (2019). 'A new linear theory of light and matter', *Journal of Physics: Conference Series* 1251, 012050. 10 p.

Williamson, J.G. and M.B. van der Mark (1997). 'Is the electron a photon with toroidal topology?' *Annales de la Fondation Louis de Broglie*, **22**(2): 133-158.

Wilson, W.O., C.G. Mayo and J.W. Head (1962). 'The fundamental importance of the Heaviside operational calculus', *Journal of the British Institution of Radio Engineers*, **24**(6): 461-477.

Winand, P.A.N. (1892). 'On polyphased currents', *Journal of the Franklin Institute*, **134**(4): 312-330.

Windred, G. (1929). 'History of the theory of imaginary and complex quantities', *The Mathematical Gazette*, **14**(203): 533-541.

Wisnesky, R.J. (2004). 'The forgotten quaternions' [pdf document from Philosophy 300 course, Stanford University] http://citeseerx.ist.psu. edu/viewdoc/download?rep=rep1&type=pdf&doi=10.1.1.168.3780.

Witten, E. (2004). 'When symmetry breaks down', *Nature*, **429**(6991): 507-508.

Wood, A.J. and Wollenberg, B.F. (1984). *Power Generation, Operation, and Control.* New York: Wiley.

Wußing, H. (2009). *6000 Jahre Mathematik: Eine kulturgeschichtliche Zeitreise*, vol. 2: *Von Euler bis zur Gegenwart.* Berlin: Springer.

Xambó-Descamps, S. (2018). *Real Spinorial Groups: A Short Mathematical Introduction.* Cham, Switzerland: Springer.

Yaglom, I.M. (1968). *Complex Numbers in Geometry.* New York: Academic Press.

Yamada, H. and H. Kanai (1982). 'Impedance calculation of solid-salient pole machine at asynchronous operation', *Electric Machines & Power Systems*, **7**(4): 231-241.

Yamamura, S. (1982). *Spiral Vector Theory of AC Circuits and Machines.* Oxford: Clarendon Press.

Yavetz, I. (1995). *From Obscurity to Enigma: The Work of Oliver Heaviside.* Basel: Birkhäuser.

Yavetz, I. (2005). 'Oliver Heaviside, electrical papers (1892)'. In: *Landmark Writings in Western Mathematics, 1640-1940*, (Eds.). I. Grattan-Guinness *et al.*, 639-652. Elsevier Science.

Yosida, K. (1980). *Functional Analysis.* Berlin: Springer.

Yosida, K. (1984). *Operational Calculus: A Theory of Hyperfunctions*, Applied Mathematical Sciences, 55. New York: Springer.

Zaborszky, J. and J.W. Rittenhouse (1954). *Electric Power Transmission: The Power System in the Steady State.* New York: Ronald Press.

Zaddach, A. (1994). *Grassmanns Algebra in der Geometrie.* Mannheim: Bibliographisches Institut & F.A. Brockhaus A.G.

Zangwill, A. (2012). *Modern Electrodynamics*. Cambridge University Press.
Zeidler, E. (2011). 'Electrical circuits as a paradigm in homology and cohomology'. In: *Quantum Field Theory, III: Gauge Theory*, (Ed.). E. Zeidler, 1009-1026. Berlin – Heidelberg: Springer.
Zeuner, G. (1884). *Treatise on Valve-Gears, with Special Consideration of the Locomotive Engines*. New York: E. & F.N. Spon.
Zhang, K. and D. Li. (2008). *Electromagnetic Theory for Microwaves and Optoelectronics*. Heidelberg: Springer.

Index

Milton Keynes UK
Ingram Content Group UK Ltd.
UKHW040053071024
449327UK00019B/522